FOUNDATIONS OF NANOTECHNOLOGY

VOLUME 3

MECHANICS OF CARBON NANOTUBES

FOUNDATIONS OF NANOTECHNOLOGY
VOLUME 3
MECHANICS OF CARBON NANOTUBES

Saeedeh Rafiei

APPLE ACADEMIC PRESS

Apple Academic Press Inc.	Apple Academic Press Inc.
3333 Mistwell Crescent	9 Spinnaker Way
Oakville, ON L6L 0A2	Waretown, NJ 08758
Canada	USA

©2015 by Apple Academic Press, Inc.

First issued in paperback 2021

Exclusive worldwide distribution by CRC Press, a member of Taylor & Francis Group
No claim to original U.S. Government works

ISBN 13: 978-1-77463-106-5 (pbk)
ISBN 13: 978-1-77188-076-3 (hbk)

Library of Congress Control Number: 2015938274

Library and Archives Canada Cataloguing in Publication

Foundations of nanotechnology.
(AAP research notes on nanoscience & nanotechnology book series)
Contents: Volume 3. Mechanics of carbon nanotubes / Saeedeh Rafiei.

Includes bibliographical references and index.
ISBN 978-1-77188-076-3 (v. 3 : bound)
1. Nanotechnology. I. Series: AAP research notes on nanoscience & nanotechnology book series

---

| T174.7.F69 2014 | 620'.5 | C2014-905376-2 |

Apple Academic Press also publishes its books in a variety of electronic formats. Some content that appears in print may not be available in electronic format. For information about Apple Academic Press products, visit our website at **www.appleacademicpress.com** and the CRC Press website at **www.crcpress.com**

ABOUT AAP RESEARCH NOTES ON NANOSCIENCE & NANOTECHNOLOGY

AAP Research Notes on Nanoscience & Nanotechnology reports on research development in the field of nanoscience and nanotechnology for academic institutes and industrial sectors interested in advanced research.

BOOKS IN THE AAP RESEARCH NOTES ON NANOSCIENCE & NANOTECHNOLOGY BOOK SERIES

Nanostructure, Nanosystems and Nanostructured Materials: Theory, Production, and Development
Editors: P. M. Sivakumar, PhD, Vladimir I. Kodolov, DSc, Gennady E. Zaikov, DSc, and A. K. Haghi, PhD

Nanostructures, Nanomaterials, and Nanotechnologies to Nanoindustry
Editors: Vladimir I. Kodolov, DSc, Gennady E. Zaikov, DSc, and A. K. Haghi, PhD

Foundations of Nanotechnology:
Volume 1: Pore Size in Carbon-Based Nano-Adsorbents
A. K. Haghi, PhD, Sabu Thomas, PhD, and Moein MehdiPour MirMahaleh

Foundations of Nanotechnology:
Volume 2: Nanoelements Formation and Interaction
Sabu Thomas, PhD, Saeedeh Rafiei, Shima Maghsoodlou, and Arezo Afzali

Foundations of Nanotechnology:
Volume 3: Mechanics of Carbon Nanotubes
Saeedeh Rafiei

ABOUT THE AUTHOR

Saeedeh Rafiei

Saeedeh Rafiei is a professional textile engineer and is currently a Research Scholar at Scuola Internazionale Superiore di Studi Avanzati, Trieste, Italy. She earned a BSc and MSc in Textile Engineering and has published several papers in journals and international conferences.

CONTENTS

LIST OF ABBREVIATIONS

AACVD	Aerosol Assisted Chemical Vapor Deposition
AAO	Anodized Aluminum Oxide
ABC	Atomistic-Based Continuum
ACFs	Activated Carbon Fibers
ACs	Activated Carbons
AFM	Atomic-Force Microscope
APCVD	Atmospheric-Pressure Chemical Vapor Deposition
ASFEM	Atomic-Scale Finite Element Method
BCC	Body-Centered Cubic
BD	Brownian Dynamics
BOA	Born-Oppenheimer Approximation
CBR	Cauchy-Born Rule
C–C	Carbon–Carbon
CCVD	Catalytic Chemical Vapor Deposition
CG	Conjugate Gradient
CHC	Cahn–Hilliard–Cook
CNTs	Carbon Nanotubes
CRS	Compressed Row Storage
CVD	Chemical Vapor Deposition
DC-PECVD	Direct Current –PECVD
DFT	Density Functional Theory
DGEBA	Diglycidyl Ether of Bisphenol A
DMTA	Dynamic Mechanical Thermal Analyzer
DPD	Dissipative Particle Dynamics
ECBR	Exponential Cauchy-Born Rule
EDA	Electron-Donor-Acceptor
EMI	Electro Magnetic Induction
ERM	Effective Reinforcing Modulus
FCC	Face-Centered Cubic
FEM	Finite Element Method
FF	Force Field
FH	Ferrum–Hydrogen

GAC	Granular Activated Carbon
GDEs	Geodesic Differential Equations
GNFs	Graphite Nano fibers
HFCVD	Hot-Filament Chemical Vapor Deposition
H–T	Halpin–Tsai
ISS	Interfacial Shear Strength
LB	Lattice Boltzmann
LJ	Lennard-Jones
MC	Monte Carlo
MD	Molecular Dynamics
MH	Multi-scale Homogenization
MM	Molecular Mechanics
MM	Molecular Mechanics
MPECVD	Microwave Plasma Chemical Vapor Deposition
MWNT	Multi-Walled Carbon Nano Tube
MWPECVD	Microwave Plasma Enhanced Chemical Vapor Deposition
NM	Newtonian Mechanics
ODEs	Ordinary Differential Equations
OLED	Organic Light Emitting Diodes
PAC	Powdered Activated Carbon
PB	Prussian Blue
PCB	Printed Circuit Board
PE	Plasma Enhanced
PECVD	Plasma Enhanced Chemical Vapor Deposition
PLA	Polylactic Acid
PLD	Pulsed Laser Deposition
PMMA	Poly (Methyl Methacrylate)
PPV	Phenylenevinylene
PSNT	Polystyrene Target
QM	Quantum Mechanics
QM	Quantum Mechanics
RF-CVD	Radio Frequency Chemical Vapor Deposition
RVE	Representative Volume Element
RVE	Representative Volume Element
SD	Steepest Descent
SOCs	Synthetic Organic Compounds
SUSHI	Simulation Utilities for Soft and Hard Interfaces
SWNT	Single-Walled Carbon Nanotube

TB	Tight Binding
TDGL	Time-Dependent Ginsburg–Landau
TETA	TriethyleneTetramine
TPD	Temperature-Programmed Desorption
T–W	Tandon–Weng
vdW	van der Waals
VGCF	Vapor Grown Carbon Fiber

LIST OF SYMBOLS

A	set of all the atoms of the sheet
a	translational period of group L
a_1 and a_2	hexagonal lattice
B	body force per unit undeformed area
B	set of all the binary bonds between pairs of adjacent atoms
C	set of all the ordered couples of adjacent bonds
C	stiffness tensor
D	dissociation energy
E	function of spectroscopic constants
F	force vector
F	force applied to the cross-sectional area
H(i) and H(j)	Hamiltonian associated with the original and new configuration
n, m	number of steps along the unit vectors
P	non-equilibrium force vector
Pi	momentum of particle i
Q	empirical dielectric constant
S	average compliance
T	torque acting at the end of an SWNT
T	total torque applied on the nanotubes
Vnb	continuous Van der Waals energy double density

Greek Symbols

α	rotational angle at ends of beam
$A_0 = a_{cc}$	equilibrium bond length
B and B_0	Euclidean bases
B_X	ball centered at X with a radius that is function of potential cut-off radius
$B_{b[\psi]}$	vector related to the bond in Ω
d_g	diameter at the energy ground
Δr, $\Delta \theta$ and $\Delta \phi$	bond stretching increment

ΔL	axial stretching deformation
Δl	difference in length after the load
ΔU	change in the sum of the mixing energy and the chemical potential of the mixture
$\Delta \beta$	relative rotation between the ends of the beam
$\varepsilon_{\alpha\beta}$	mid-surface strains
Em	Young's modulus of the filler
Ef	Young's modulus of matrix
ε	predefined tolerance
$\vec{F}_i(t)$	force acting on the ith atom
F_{ij}^c	conservative force of particle j acting on particle i, γ and σ are constants
H_0	initial length
J_0	cross-sectional polar inertia of the SWNT
k_r, k_θ and k_τ	bond stretching force constant
k_l and k_p	stiffness coefficients
KB	Boltzmann constant
L_0	length of the tube
l_0	length on graphene sheet
l_g	length at energy ground for the tube
r_0	carbon-carbon distance
r_{ij}	distance between the atoms i and j
$\vec{r}_i$	atomic position
θ_{ijk}	angle between bonds $i-j$ and $i-k$,
θ	angle of twist
v_f	volume fraction of filler
v_0	Poisson's ratio of the matrix
U^a, U^b and U^c	energies associated with truss elements
Uv	Hookian potential energy
U^{vdw}	covalent bond stretching
U_r	bond stretch interaction
U_θ	bond angle bending
U_ϕ	dihedral angle torsion
U_ω	improper (out-of-plane) torsion
U_{vdw}	non-bonded van der Waals interaction
Uv	Hookian potential energy
U_r and U_A	stretching energy
U_θ and U_M	bending energy

U_τ and U_T	torsional energy
V_R and V_A	repulsive and attractive pair terms
$Vp(r)$	potential function for bend stretch
vel_a	velocity function of the atom a • A
$\psi^{(0)}$	initial guess of equilibrium state
$\psi = \psi_e$	harmonic oscillator component
ψ_0	collision diameter
ν	Poisson ratio
ϕ	stands for double contraction of tensors
$\phi(m)$	Euler function
ε	dislocation energy
ζ_f	shape parameter depending on filler geometry
$(-\gamma P)$	dissipative
$(\sigma \zeta(t))$	random force terms
$\zeta(t)$	Gaussian random noise term
$\sigma(x)$	shape parameter
$\sigma(x)$	stress field
σ_v	vertical mirror plane
Λ	vibrational quantum number

PREFACE

In this book, the modeling of mechanical properties of CNT/polymer nanocomposites is reviewed. The book starts with the structural and intrinsic mechanical properties of CNTs. Then we introduce some computational methods that have been applied to polymer nanocomposites, covering from molecular scale (e.g., molecular dynamics, Monte Carlo), micro scale (e.g., Brownian dynamics, dissipative particle dynamics, lattice Boltzmann, time-dependent Ginzburg-Landau method, dynamic density functional theory method) to mesoscale and macroscale (e.g., micromechanics, equivalent-continuum and self-similar approaches, finite element method). Hence, the knowledge and understanding of the nature and mechanics of length and orientation of nanotube and load transfer between nanotube and polymer are critical for the manufacturing of enhanced carbon nanotube-polymer composites and will enable the tailoring of the interface for specific applications or superior mechanical properties. In this book a state of these parameters in mechanics of carbon nanotube polymer composites is discussed along with some directions for future research in this field.

CHAPTER 1

INTRODUCTION

CONTENTS

1.1   INTRODUCTION

Carbon nanotubes were first observed by Iijima [1], almost two decades ago, andsince then, extensive work has been carried out to characterize their properties [2–4]. A wide range of characteristic parameters has been reported for carbon nanotubes nanocomposite. There are contradictory reports that show the influence of carbon nanotubes on a particular property (e.g., Young's modulus) to be improving, indifferent or even deteriorating [5]. However, from the experimental point of view, it is a great challenge to characterize the structure and to manipulate the fabrication of polymer nanocomposite. The development of such materials is still largely empirical and a finer degree of control of their properties cannot be achieved so far. Therefore, computer modeling and simulation will play an ever increasing role in predicting and designing material properties, and guiding such experimental work as synthesis and characterization, For polymer nanocomposite, computer modeling and simulation are especially useful in the hierarchical characteristics of the structure and dynamics of polymer nanocomposite ranging from molecular scale, micro scale to mesoscale and macroscale, in particular, the molecular structures and dynamics at the interface between nanoparticles and polymer matrix.The purpose of this review is to discuss the application of modeling and simulation techniques to polymer nanocomposite. This includes a broad subject covering methodologies at various length and time scales and many aspects of polymer nano composites. We organize the review as follows. In Section 1.1 we will discuss about Carbon nanotubes's (CNTs) and nano composite properties. In Section 1.2, we introduce briefly the computational methods used so far for the systems of polymer nanocomposite which can be roughly divided into three types: molecular scale methods (e.g., molecular dynamics (MD), Monte Carlo (MC)), micro scale methods (e.g., Brownian dynamics (BD), dissipative particle dynamics (DPD), lattice Boltzmann (LB), time dependent Ginzburg-Lanau method, dynamic density functional theory (DFT) method), and mesoscale and macroscale methods (e.g., micromechanics, equivalent-continuum and self-similar approaches, finite element method (FEM)) [6]. Many researchers used this method for determine the mechanical properties of nanocomposite that in Section 1.3 will be discussed. In Section 1.4 modeling of interfacial load transfer between CNT and polymer in nanocomposite will be introduced and finally we conclude the review by emphasizing the current challenges and future research directions.

1.2 CARBON NANOTUBES' (CNTs) AND NANO COMPOSITE PROPERTIES

CNTs are one dimensional carbon materials with aspect ratio greater than 1000. They are cylinders composed of rolled-up graphite planes with diameters in nanometer scale [1, 7, 8]. The cylindrical nano tube usually has at least one end capped with a hemisphere of fullerene structure. Depending on the process for CNT fabrication, there are two types of CNTs [7–10]: single-walled CNTs (SWCNTs) and multi walled CNTs (MWCNTs). SWCNTs consist of a single graphene layer rolled up into a seamless cylinder whereas MWCNTs consist of two or more concentric cylindrical shells of graphene sheets coaxially arranged around a central hollow core with Vander Waals forces between adjacent layers. According to the rolling angle of the graphene sheet, CNTs have three chiralities: armchair, zigzag and chiral one. The tube chirality is defined by the chiral vector, $C_h = na_1 + ma_2$(Fig. 1.1), where the integers (n, m) are the number of steps along the unit vectors (a_1 and a_2) of the hexagonal lattice [11, 12]. Using this (n, m) naming scheme, the three types of orientation of the carbon atoms around the nano tube circumference are specified. If n = m, the nanotubes are called "armchair." If m = 0, the nanotubes are called "zigzag." Otherwise, they are called "chiral." The chirality of nanotubes has significant impact on their transport properties, particularly the electronic properties. For a given (n, m) nano tube, if (2n + m) is a multiple of 3, then the nano tube is metallic, otherwise the nano tube is a semiconductor. Each MWCNT contains a multilayer of graphene, and each layer can have different chirality, so the prediction of its physical properties is more complicated than that of SWCNT. Figure 1.1 shows the CNT with different chiralities.

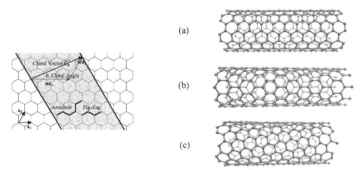

FIGURE 1.1. Schematic diagram showing how a hexagonal sheet of graphene is rolled to form a CNT with different chiralities (A: armchair; B: zigzag; C: chiral).

1.3   CLASSIFICATION OF CNT/POLYMER NANOCOMPOSITES

Polymer composites, consisting of additives and polymer matrices, including thermoplastics, thermo sets and elastomers, are considered to be an important group of relatively inexpensive materials for many engineering applications. Two or more materials are combined to produce composites that possess properties that are unique and cannot be obtained each material acting alone. For example, high modulus carbon fibers or silica particles are added into a polymer to produce reinforced polymer composites that exhibit significantly enhanced mechanical properties including strength, modulus and fracture toughness. However, there are some bottlenecks in optimizing the properties of polymer composites by employing traditional micron-scale fillers. The conventional filler content in polymer composites is generally in the range of 10–70 wt.%, which in turn results in a composite with a high density and high material cost. In addition, the modulus and strength of composites are often traded for high fracture toughness [13]. Unlike traditional polymer composites containing micron-scale fillers, the incorporation of nanoscale CNTs into a polymer system results in very short distance between the fillers, thus the properties of composites can be largely modified even at an extremely low content of filler. For example, the electrical conductivity of CNT/epoxy nano composites can be enhanced several orders of magnitude with less than 0.5 wt.% of CNTs [14]. As described previously, CNTs are among the strongest and stiffest fibers ever known. These excellent mechanical properties combined with other physical properties of CNTs exemplify huge potential applications of CNT/polymer nanocomposite. Ongoing experimental works in this area have shown some exciting results, although the much-anticipated commercial success has yet to be realized in the years ahead. In addition, CNT/polymer nanocomposite are one of the most studied systems because of the fact that polymer matrix can be easily fabricated without damaging CNTs based on conventional manufacturing techniques, a potential advantage of reduced cost for mass production of nanocomposite in the future. Following the first report on the preparation of a CNT/polymer nanocomposite in 1994 [15], many research efforts have been made to understand their structure-property relationship and find useful applications in different fields, and these efforts have become more pronounced after the realization of CNT fabrication in industrial scale with lower costs in the beginning of the twenty-first century. According to the specific application, CNT/polymer nano composites can be classified as structural or functional composites [16]. For the structural composites, the unique mechanical properties of CNTs, such as the high modulus, tensile strength and strain to fracture, are explored to obtain structural materials with much improved mechanical properties. As for CNT/polymer functional composites many other unique properties of CNTs, such as electrical, thermal, optical and damping properties along with their excellent mechanical properties, are used to develop multifunctional composites for applications in the fields of heat resistance, chemical sensing, electrical and thermal management, photoemission, electromagnetic absorbing and energy storage performances, etc.

1.4   MOLECULAR STRUCTURE OF CNTs

1.4.1   BONDING MECHANISMS

The mechanical properties of CNTs are closely related to the nature of the bonds between the carbon atoms. The bonding mechanism in a carbon nanotubes system is similar to that of graphite, since a CNT can be thought of as a rolled-up graphene sheet. The atomic number for carbon is 6, and the atom electronic structure is $1s^2 2s^2 2p^2$ in atomic physics notation. For a detailed description of the notation and the structure, readers may refer to basic textbooks on general chemistry or physics [17].

When carbon atoms combine to form graphite, sp^2 hybridization occurs. In this process, one s-orbital and two p-orbitals combine to form three hybrid sp^2-orbitals at 120° to each other within a plane (Fig. 1.1). This in-plane bond is referred to as an σ-bond (*sigma*–bond). This is a strong covalent bond that binds the atoms in the plane, and results in the high stiffness and high strength of a CNT. The remaining π-orbital is perpendicular to the plane of the σ-bonds. It contributes mainly to the interlayer interaction and is called the π-bond (*pi*–bond). These out-of-planes, delocalized p-bonds interact with the p-bonds on the neighboring layer. This interlayer interaction of atom pairs on neighboring layers is much weaker than s-bond. For instance, in the experimental study of *shell sliding* [18, 19] it was found that the shear strength between the outermost shell and the neighboring inner shell was 0.08 MPa and 0.3 MPa according to two separate measurements on two different MWCNTs. The bond structure of a graphene sheet is shown in Fig. 1.1.

1.4.2 FROM GRAPHENE SHEET TO SINGLE-WALLED NANOTUBES

There are various ways of defining a unique structure for each carbon nanotubes. One way is to think of each CNT as a result of rolling a graphene sheet, by specifying the direction of rolling and the circumference of the cross-section. Shown in Fig. 1.2 is a graphene sheet with defined roll-up vector *r*. After rolling to form a NT, the two end nodes coincide. The notation we use here is adapted from [2, 8, 20].

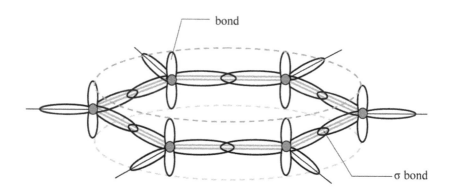

FIGURE 1.2   Basic hexagonal bonding structure for one graphite layer (the graphene sheet); carbon nuclei shown as filled circles, out-of-plane π-bonds represented as delocalized (dotted line), and σ-bonds connect the C nuclei in-plane.

$$r = na + mb \tag{1}$$

Note that *r*(bold solid line in Fig. 1.2) can be expressed as a linear combination of base vectors *a* and *b* (dashed line in Fig. 1.2) of thewith *n* and *m* being integers. Different types of NT are thus uniquely defined by of the values of *n* and *m* and the ends are closed with caps for certain types of fullerenes. Three major categories of NT can also be defined based on the chiral angle u (Fig. 1.2) as follows

$$\theta = 0 \, "Zigzag"$$

$$0 < \theta < 30, "Chiral \tag{2}$$

$$\theta = 30, "'ArmChair"$$

Based on simple geometry, the diameter d and the chiral angle θ of the NT can be given as

$$d = 0.783\sqrt{n^2 + nm + m^2} \, A \tag{3}$$

$$\theta = \sin^{-1}\sin^{-1}\left[\frac{\sqrt{3}m}{2(n^2 + nm + m^2)}\right] \tag{4}$$

Most CNTs to date have been synthesized with closed ends. Fujita et al. [21] and Dresselhaus et al. [2, 22] have shown that NTs, which are larger than (5, 5) and (9, 0) tubes can be capped. Based on Euler's theorem of Polyhedra [23], which relates the numbers of the edges, faces and vertices, along with additional knowledge of the minimum energy structure of fullerenes, they conclude that any cap must contain 6 pentagons that are isolated from each other (Figs. 1.3 and 1.4).

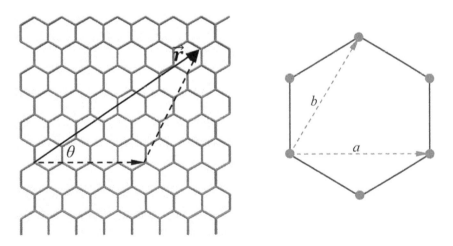

FIGURE 1.3   Definition of roll up vectors as linear combinations of base vector a and b.

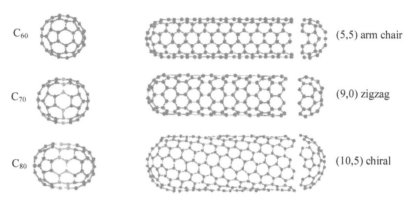

FIGURE 1.4 Examples of zigzag, chiral, and arm-chair nanotubes and their caps corresponding to different types of fullerenes.

For NTs with large radius, there are different possibilities of forming caps that satisfy this requirement. The experimental results of Iijima et al. [24] and Dravid [25] indicate a number of ways that regular-shaped caps can be formed for large diameter tubes. Bill-like [26] and semitoroidal [24] types of termination have also been reported. Experimental observation of CNTs with open ends can be found in [24].

1.4.3 MULTI-WALLED CARBON NANOTUBES AND SCROLL-LIKE STRUCTURES

The first carbon nanotubes discovered [1] were multi-walled carbon nanotubes. Transmission electron microscopy studies on MWCNTs suggest a Russian doll-like structure (nested shells) and give interlayer spacing of approximately; 0.34 nm [27, 28] close to the interlayer separation of graphite, 0.335 nm. However, Kiang et al. [29] have shown that the interlayer spacing for MWCNTs can range from 0.342 to 0.375 nm, depending on the diameter and number of nested shells in the MWCNT. The increase in inter shell spacing with decreased nano tube diameter is attributed to the increased repulsive force as a result of the high curvature. The experiments by Zhou et al. [28], Amelincx et al. [30] and Lavin et al. [31] suggested an alternative "scroll" structure for some MWCNTs, like a cinnamon roll. In fact, both forms might be present along a given MWCNT and separated by certain types of defects. The energetic analysis by Lavin et al. [31] suggests the formation of a scroll, whichmay then convert into a stablemulti-wall structure composed of nested cylinders.

1.5   STRUCTURAL CHARACTERISTICS OF CARBON NANOTUBES

Single-walled carbon nanotubes (SWNT) can be viewed as a graphene sheet that has been rolled into atube. A multi-walled carbon nanotubes (MWNT) is composed of concentric graphitic cylinders with closedcaps at both ends and the graphitic layer spacing is about 0.34 nm. Unlike diamond, whichassumes a 3Dcrystal structure with each carbon atom having four nearest neighbors arranged in a tetrahedron, graphiteassumes the form of a 2D sheet of carbon atoms arranged in a hexagonal array. In this case, each carbonatom has three nearest neighbors.

The atomic structure of nanotubes can be described in terms of the tube chirality, or helicity,which isdefined by the chiral vector C_h and the chiral angle θ. In Fig. 1.5, we can visualize cutting the graphite sheetalong the dotted lines and rolling the tube so that the tip of the chiral vector touches its tail. The chiralvector, also known as the roll-up vector, can be described by the following equation:

$$\vec{C_h} = n\vec{a_1} + m\vec{a_2},$$

where the integers (n; m) are the number of steps along the zigzag carbon bonds of the hexagonal lattice anda$_1$ and a$_2$ are unit vectors. The chiral angle determines the amount of 'twist' in the tube. The chiral anglesare $0°$ and $30°$ for the two limiting cases which are referred to as zigzag and armchair, respectively (Fig. 1.6).

In terms of the roll-up vector, the zigzag nanotubes is denoted by (n, 0) and the armchair nanotubes (n, n).The roll-up vector of the nanotubes also defines the nanotubes diameter.

The physical properties of carbon nanotubes are sensitive to their diameter, length and chirality.

In particular, tube chirality is known to have a strong influence on the electronic properties of carbonnanotubes. Graphite is considered to be a semimetal, but it has been shown that nanotubes can be eithermetallic or semiconducting, depending on tube chirality [2]. The influence of chiralityon the mechanical properties of carbon nanotubes has also been reported [32, 33].

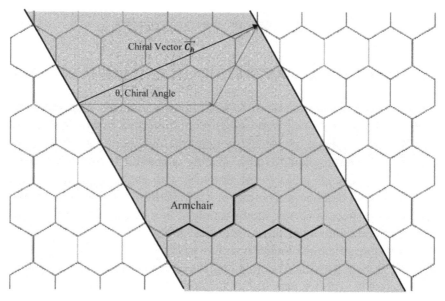

FIGURE 1.5   Schematic diagrams of a hexagonal grapheme sheet.

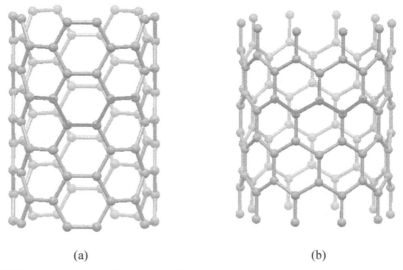

(a)   (b)

FIGURE 1.6   Schematic diagrams of (a) an armchair and (b) A zigzag nanotubes.

1.6 CHARACTERIZATION OF CARBON NANOTUBES

Significant challenges exist in both the micromechanical characterization of nanotubes and the modeling of the elastic and fracture behavior at the nano-scale. Challenges in characterization of nanotubes and their composites include (a) complete lack of micromechanical characterization techniques for direct property measurement, (b) tremendous limitations on specimen size, (c) uncertainty in data obtained from indirect measurements, and(d) inadequacy in test specimen preparation techniques and lack of control in nanotubes alignment and distribution.

In order better to understand the mechanical properties of carbon nanotubes, a number of investigators have attempted to characterize carbon nanotubes directly. Treacy et al. [34] first investigated the elastic modulus of isolated multi-walled nanotubes by measuring, in the transmission electron microscope, the amplitude of their intrinsic thermal vibration. The average value obtained over 11 samples was 1.8 TPa. Direct measurement of the stiffness and strength of individual, structurally isolated multi-wall carbon nanotubes has been made with an atomic-force microscope (AFM). Wong et al. [35] were the first to perform direct measurement of the stiffness and strength of individual, structurally isolated multi-wall carbon nanotubes using atomic force microscopy. The nano tube was pinned at one end to molybdenum disulfide surfaces and load was applied to the tube by means of the AFM tip. The bending force was measured as a function of displacement along the unpinned length, and a value of 1.26 TPa was obtained for the elastic modulus. The average bending strength measured was 14.2–8GPa.

Single-walled nanotubes tend to assemble in 'ropes' of nanotubes. Salvetat et al. [36] measured the properties of these nano tube bundles with the AFM. As the diameter of the tube bundles increases, the axial and shear moduli decrease significantly. This suggests slipping of the nanotubes within the bundle. Walters et al. [37] further investigated the elastic strain of nanotubes bundles with the AFM. On the basis of their experimental strain measurements and an assumed elastic modulus of 1.25 TPa, they calculated yield strength of 45±7 GPa for the nanotubes ropes. Indeed, their calculated value for strength would be much lower if the elastic modulus of the nanotubes bundle is decreased as a consequence of slipping within the bundle, suggested by Salvetat et al. [36]. Yu et al. [38, 39] have investigated the tensile loading of multi-walled nanotubes and single-walled nanotubes ropes. In their work, the nanotubes were attached between two opposing AFM tips and loaded under tension. For multi-walled carbon nanotubes [39] the failure of the outermost tube occurred followed by pullout of the inner nanotubes. The experimentally calculated tensile strengths of the outermost layer ranged from 11 to 63 GPa and the elastic modulus ranged from 270 to 950 GPa. In their subsequent investigation of single-walled nanotubes ropes [40], they assumed that only the outermost tubes assembled in the rope carried the load during the experiment, and they calculated tensile strengths of 13 to 52 GPa and average elastic moduli of 320 to 1470 GPa. Xie et al. [41] also tested ropes of multi-walled nanotubes in tension. In their experiments, the obtained tensile strength and modulus were 3.6 and 450 GPa, respectively. It was suggested that the lower values for strength and stiffness may be a consequence of defects in the CVD-grown nanotubes.

1.7 MECHANICS OF CARBON NANOTUBES

As discussed in the previous section, nanotubes deformation has been examined experimentally. Recent investigations have shown that carbon nanotubes possess remarkable mechanical properties, such as exceptionally high elastic modulus [42, 43], large elastic strain and fracture strain sustaining capability [44, 45]. Similar conclusions have also been reached through some theoretical studies [46–49], although very few correlations between theoretical predictions and experimental studies have been made. In this section, we examine the mechanics of both single walled and multi-walled nanotubes.

1.7.1 SINGLE-WALLED CARBON NANOTUBES

Theoretical studies concerning the mechanical properties of single-walled nanotubes have been pursued extensively. Overney et al. [46] studied the low-frequency Vibrational modes and structural rigidity of long nanotubes consisting of 100, 200 and 400 atoms. The calculations were based on an empirical Keating Hamiltonian with parameters determined from first principles. A comparison of the bending stiff nesses of single-walled nanotubes and an iridium beam was presented.

The bending stiffness of the iridium beam was deduced by using the continuum Bernoulli-Euler theory of beam bending. Overney et al. [46] concluded that the beam bending rigidity of a nano tube exceeds the highest values found in any other presently available materials. Besides their experimental observations, Iijima et al. [50] examined response of nanotubes under compression using molecular dynamics simulations. They simulated the deformation properties of single and multi-walled nanotubes bent to large angles. Their experimental and theoretical results show that nanotubes are remarkably flexible. The bending is completely reversible up to angles in excess of 110°, despite the formation of complex kink shapes.

Ru [51] noticed that actual bending stiffness of single walled nanotubes is much lower than that given by the elastic-continuum shell model if the commonly defined representative thickness is used. Ru [51] proposed the use of an effective nanotubes bending stiffness as a material parameter not related to the representative thickness. With the aid of this concept, the elastic shell equations can be readily modified and then applied to single-walled nanotubes. The computational results based on this concept show a good agreement with the results from MD simulations.

Vaccarini et al. [52] investigated the influence of nanotubes structure and chirality on the elastic properties in tension, bending, and torsion. They found that the chirality played a small influence on the nanotubes tensile modulus. However, the chiral tubes exhibit asymmetric Torsional behavior with respect to left and right twist, whereas the armchair and zig-zag tubes do not exhibit this asymmetric Torsional behavior. A relatively comprehensive study of the elastic properties of single-walled nanotubes was reported by Lu [47]. In this study, Lu [53] adopted an empirical lattice dynamics model, which has been successfully adopted in calculating the phonon spectrum and elastic properties of graphite. In this lattice-dynamics model, atomic interactions in a single carbon layer are approximated by a sum of pair-wise harmonic potentials between atoms. The local structure of a nanotubes layer is constructed from conformal mapping of a graphite sheet on to a cylindrical surface. Lu's work attempted to answer such basic questions as: (a) how do elastic properties of nanotubes depend on the structural details, such as size and chirality? And (b) how do elastic properties of nanotubes compare with those of graphite and diamond? Lu concluded that the elastic properties of nanotubes are insensitive to size and chirality. The predicted Young's modulus (~1 TPa), shear modulus

(~0.45 TPa), and bulk modulus (~0.74 TPa) are comparable to those of diamond. Hernandez et al. [51] performed calculations similar to those of Lu and found slightly higher values (~1.24 TPa) for the Young's moduli of tubes. But unlike Lu, they found that elastic moduli are sensitive to both tube diameter and structure. Besides their unique elastic properties, the inelastic behavior of nanotubes has also received considerable attention. Yakobson et al. [12, 49] examined the instability behavior of carbon nanotubes beyond linear response by using a realistic many-body Ters off-Brenner potential and MD simulations. Their MD simulations show that carbon nanotubes, when subjected to large deformations, reversibly switch into different morphological patterns. Each shape change corresponds to an abrupt release of energy and a singularity in the stress/strain curve. These transformations are explained well by a continuum shell model. With properly chosen parameters, their model provided a very accurate 'roadmap' of nanotubes behavior beyond the linear elastic regime. They also made MD simulations to single- and double-walled nanotubes of different chirality and at different temperatures [48]. Their simulations show that nanotubes have an extremely large breaking strain (in the range 30–40%) and the breaking strain decreases with temperature. Yakobson [54] also applied dislocation theory to carbon nanotubes for describing their main routes of mechanical relaxation under tension. It was concluded that the yield strength of a nanotubes depends on its symmetry and it was believed that there exists an intra-molecular plastic flow. Under high stress, this plastic flow corresponds to a motion of dislocations along helical paths within the nanotubes wall and causes a stepwise necking, a well-defined new symmetry, as the domains of different chiral symmetry are formed. As a result, both the mechanical and electronic properties of carbon nanotubes are changed.

The single walled nanotubes produced by laser ablation and arc-discharge techniques have a greater tendency to form 'ropes' or aligned bundles [55, 56]. Thus, theoretical studies have been made to investigate the mechanical properties of these nanotubes bundles. Ru [57] presented a modified elastic-honeycomb model to study elastic buckling of nanotubes ropes under high pressure. Ru gave a simple formula for the critical pressure as a function of nanotubes Young's modulus and wall thickness-to-radius ratio. It was concluded that single-walled ropes are susceptible to elastic buckling under high pressure and elastic buckling is responsible for the pressure-induced abnormalities of vibration modes and electrical resistivity of single walled nanotubes. Popov et al. [32] studied the elastic properties of triangular crystal lattices formed by single-walled nanotubes by using analytical expressions based on a force constant lattice dynamics model [58]. They calculated various elastic constants of nanotubes crystals for nanotubes types, such as armchair and zigzag. It was shown that the elastic modulus, Poisson's ratio and bulk modulus clearly exhibit strong dependence on the tube radius. The bulk modulus was found to have a maximum value of 38 GPa for crystals composed of single walled nanotubes with 0.6 nm radius.

1.7.2   MULTI-WALLED CARBON NANOTUBES

Multi-walled nanotubes are composed of a number of concentric single walled nanotubes held together with relatively weak Vander Waals forces. The multilayered structure of these nanotubes further complicates the modeling of their properties.

Ruoff and Lorents [59] derived the tensile and bending stiffness constants of ideal multi-walled nanotubes in terms of the known elastic properties of graphite. It is suggested that unlike the strongly anisotropic thermal expansion in conventional carbon fibers and graphite, the thermal expansion of carbon nanotubes is essentially isotropic. However, the thermal conductivity of nanotubes is believed to be highly anisotropic and its magnitude along the axial direction is perhaps higher than that of any other material. Lu [47] also calculated the elastic properties of many multi-walled nanotubes formed by single-layer tubes by means of the empirical-lattice dynamics model. It was found that elastic properties are insensitive to different combinations of parameters, such as chirality, tube radius and numbers of layers and the elastic properties are the same for all nanotubes with a radius larger than one nm. Interlayer Vander Waals interaction has a negligible contribution to both the tensile and shear stiffness.

Govindjee and Sackman [60] were the first to examine the use of continuum mechanics to estimate the properties of multi-walled nanotubes. They investigated the validity of the continuum approach by using Bernoulli- Euler bending to infer the Young's modulus. They used a simple elastic sheet model and showed that at the nanotubes scale the assumptions of continuum mechanics must be carefully respected in order to obtain reasonable results. They showed the explicit dependence of 'material properties' on system size when a continuum cross-section assumption was used. Ru [61] used the elastic-shell model to study the effect of Vander Waals forces on the axial buckling of a double- walled carbon nano tube. The analysis showed that the Vander Waals forces do not increase the critical axial buckling strain of a double-walled nanotubes Ru [62, 63] thereafter also proposed a multiple column model that considers the interlayer radial displacements coupled through the Vander Waals forces.

This model was used to study the effect of interlayer displacements on column buckling. It was concluded that the effect of interlayer displacements could not be neglected unless the Vander Waals forces are extremely strong.

Kolmogorov and Crespi [64] investigated the interlayer interaction in two-walled nanotubes. Registry-dependent two-body graphite potential was developed. It was demonstrated that the tightly constrained geometry of a multi-walled nano tube could produce an extremely smooth solid-solid interface wherein the corrugation against sliding does not grow with system size. The energetic barrier to interlayer sliding in defect-free nanotubes containing thousands of atoms can be comparable to that for a single unit cell of crystalline graphite. Although there is experimental variability in the direct characterization of carbon nanotubes, theoretical and experimental observations reveal their exceptional properties. As a consequence, there has been recent interest in the development of nanotubes-based composites. Although most research has focused on the development of nanotubes-based polymer composites, attempts have also been made to develop metal and ceramic-matrix composites with nanotubes as reinforcement.

1.8   NANO TUBE-BASED POLYMER COMPOSITES

The reported exceptional properties of nanotubes have motivated others to investigate experimentally the mechanics of nanotubes-based composite films. Uniform dispersion within the polymer matrix and improved nanotubes/matrix wetting and adhesion are critical issues in the processing of these nano composites. The issue of nanotubes dispersion is critical to efficient reinforcement. In the work of Salvetat et al. [36] discussed earlier, slipping of nanotubes when they are assembled in ropes significantly affects the elastic properties.

In addition to slipping of tubes that are not bonded to the matrix in a composite, the aggregates of nanotubes ropes effectively reduce the aspect ratio (length/diameter) of the reinforcement. It is, however, difficult to obtain a uniform dispersion of carbon nanotubes in the polymer matrix. Shaffer and Windle [65] were able to process carbon nanotubes/polyvinyl-alcohol composite films for mechanical characterization. The tensile elastic modulus and damping properties of the composite films were assessed in a dynamic mechanical thermal analyzer (DMTA) as a function of nanotubes loading and temperature. From the theory developed for short-fiber composites, a nanotubes elastic modulus of 150 MPa was obtained from the experimental data.

This value in a microscopic composite is well below the values reported for isolated nanotubes. It is not clear whether this result is a consequence of imperfections in the graphite layers of catalytically grown nanotubes used for the investigation or whether it relates to a fundamental difficulty in stress transfer.

Qian et al. [66] characterized carbon-nanotubes/polystyrene composites. With only the addition of 1% by weight (about 0.5% by volume) they achieved between 36–42% increase in the elastic stiffness and a 25% increase in the tensile strength.

Jia et al. [67] showed that the nanotubes can be initiated by a free radical initiator, AIBN (2, 2'-azobisisobutyronitrile), to open their p bonds. In their study of carbon-nanotubes/poly (methyl methacrylate)(PMMA) composites, the possibility exists to form a C–C bond between the nanotubes and the matrix. Gong et al. [68] investigated surfactant-assisted processing of nanotubes composites with a non-ionic surfactant. Improved dispersion and interfacial bonding of the nanotubes in an epoxy matrix resulted in a 30% increase in elastic modulus with addition of 1 wt.% nanotubes. Lordi and Yao [69] looked at the molecular mechanics of binding in nanotubes-based composites. In their work, they used force-field-based molecular-mechanics calculations to determine the binding energies and sliding frictional stresses between pristine carbon nanotubes and different polymeric matrix materials. The binding energies and frictional forces were found to play only a minor role in determining the strength of the interface. The key factor in forming a strong bond at the interface is having a helical conformation of the polymer around the nanotubes. They suggested that the strength of the interface may result from molecular-level entanglement of the two phases and forced long-range ordering of the polymer.

Because the interaction at the nanotubes/matrix interface is critical to understanding the mechanical behavior of nanotubes-based composites, a number of researchers have investigated the efficiency of interfacial stress transfer. Wagner et al. [70] examined stress-induced fragmentation of multi-walled carbon nanotubes in polymer films. Their nanotubes-containing film had a thickness of approximately 200 mm. The observed fragmentation phenomenon was attributed to either process induced stress resulting from curing of the polymer or tensile stress generated by polymer deformation and transmitted to the nanotubes. From estimated values of nanotubes axial normal stress and elastic modulus, Wagner et al. concluded that the nanotubes/polymer interfacial shear stress is on the order of 500 MPa and higher. This value, if reliable, is an order of magnitude higher than the stress-transfer ability of current advanced composites and, therefore, such interfaces are more able than either the matrix or the nanotubes themselves to sustain shear. In further work, Lourie and Wagner [71, 72] investigated tensile and compressive fracture in nanotubes-based composites.

Stress transfer has also been investigated by Raman spectroscopy. Cooper et al. [73] prepared composite specimens by applying an epoxy-resin/nanotubes mixture to the surface of an epoxy beam. After the specimens were cured, stress transfer between the polymer and the nanotubes was detected by a shift in the G. Raman band (2610 cm^{-1}) to a lower wave number. The shift in the G. Raman band corresponds to strain in the graphite structure, and the shift indicates that there is stress transfer, andhence reinforcement, by the nanotubes. It was also concluded that the effective modulus of single-walled nanotubes dispersed in a composite could be over 1 TPa and that of multi-walled nanotubes was about 0.3 TPa. In their investigation of single-walled nanotubes/epoxy composites, Ajayan et al. [74] suggest that their nearly constant value of the Raman peak in tension is related to tube sliding within the nano tube bundles and, hence, poor interfacial load transfer between the nanotubes. Similar results were obtained by Schadler et al. [75] Multi-walled nano tube/epoxy composites were tested in both tension and compression. The compressive modulus was found to be higher than the tensile modulus of the composites, and the Raman peak was found to shift only in compression, indicating poor interfacial load transfer in tension.

Even with improved dispersion and adhesion, micromechanical characterization of these composites is difficult because the distribution of the nanotubes is random. Thus, attempts have been made to align nanotubes in order better to elucidate the reinforcement mechanisms. Jin et al. [76] showed that aligned nanotubes composites could be obtained by mechanical stretching of the composite. X-ray diffraction was used to determine the orientation and degree of alignment.

Bower et al. [45] further investigated the deformation of carbon nanotubes in these aligned films. Haggen Mueller et al. [77] showed that melt spinning of single wall nanotubes in fiber form can also be used to create a well-aligned nanotubes composite.

In addition to alignment of the carbon nanotubes, researchers have attempted to spin carbon fibers from carbon nanotubes [78–80]. Andrews et al. [78] dispersed 5 wt.% single-walled nanotubes in isotropic petroleum pitch. Compared to isotropic pitch fibers without nanotubes, the tensile strength was improved by ~90%, the elastic modulus was improved by ~150% and the electrical conductivity increased by 340%. Because the pitch matrix is isotropic, the elastic modulus is 10–20 times less than that of mesophase pitch fibers used in composite materials. Further developments in this area may potentially create a new form of carbon fiber that has exceptional flexibility as well as stiffness and strength. Vigolo et al. [80] technique for spinning nanotubes based fibers involves dispersing the nanotubes in surfactant solutions followed by re condensing the nanotubes in the stream of a polymer solution to form macroscopic fibers and ribbons. Their work indicates that there is preferential orientation of the nanotubes along the axis of the ribbons. Although the elastic modulus of the nano tube fibers (9–15 GPa) is far below the values for individual nanotubes or conventional carbon fibers, the demonstrated resilience of the fibers gives hope for future improvements.

KEYWORDS

- **Carbon nanotubes (CNT)**
- **Nanocomposites**
- **Polymer**

CHAPTER 2

MODELING AND SIMULATION TECHNIQUES

CONTENTS

2.1   MODELING OF CARBON NANOTUBES BEHAVIOR

The investigations on carbon nanotubes behavior have been mainly focused on the experimental description and molecular dynamics simulations such as classical molecular dynamics, tight-binding molecular dynamics and the AB initio method. However, the researchers have been seeking more efficient computational methods with which it is possible to analyze the large scale of CNTs in a more general manner. Yakobson [12] found that the continuum shell model could predict all changes of buckling patterns in atomic molecular-dynamics simulations. The analogousness of the cylindrical shell model and CNTs leads to extensive application of the shell model for CNT structural analysis. Ru [57, 81] used the CM approach and stimulated the effect of van der Waals forces by applying a uniformly distributedpressure field on the wall, the pressure field was adjusted so as to give the same resultant force on each wall of the tube. It has verified that the mechanical responses of CNTs can be efficiently and reasonably predicted by the shell model provided that the parameters, such as Young's modulus and effective wall thickness, are judiciously adopted. Wang et al. [82] studied buckling of double-walled carbon nanotubes under axial loads, by modeling CNTs using solid shell elements. Han et al. [83, 84] investigated Torsional buckling of a DWNT and MWNT embedded in an elastic medium. Han et al. [85]also studied bending instability of double-walled carbon nanotubes. Yao and Han [84]analyzed the thermal effect on axially compressed and Torsional buckling of a multi-walled carbon nano tube. Some conclusions were drawn that at low and room temperature the critical load for infinitesimal buckling of a multi-walled carbon nanotubes increases as the value of temperature change increases, while at high temperatures the critical load for infinitesimal buckling of a multi-walled carbon nanotubes decreases as the value of temperature change increases. Nonlinear post buckling behavior of carbon nanotubes under large strain is significant to which great attention is paid by some researchers [86, 87]. The Torsional post buckling behavior of single-walled or multi-walled carbon nanotubes was discussed in details by Yiao and Han [88]The problems encountered in the numerical modeling of pristine and defective carbon nanotubes were demonstrated in details by Muc [89–91]. In the mentioned references, the linear and non-linear (iterative) approaches were illustrated.

A successful work has been also conducted with continuum modeling such as dynamic studies. A comprehensive review of the literature dealing with the analysis of wave propagation in CNTs with the use of shell theories was presented by Liew and Wang [92], althoughthe authors focused the attention mainly on the application of thick shell theories. They pointed out that there was growing interest in the terahertz physics of nanoscale materials and devices, which opened a new topic on phonon dispersion of CNTs, especially in the terahertz frequency range. Hu et al. [93]proposed to use non-local shell theories in the analysis of elastic wave propagation in single-or double-walled carbon nanotubes. However, it is worth to emphasize that the Vibrational characteristics of CNTs are studied with the use of both beam theories and different variants of shell theories (see e.g., Natsuki et al. [94] and Ghorbanpourarani et al. [95]). Recently, thermal vibrations have attracted considerable attention (see e.g., Tylikowski [96]), where the dynamic stability analysis was conducted with the use of stochastic methods. The cited above work demonstrates also another tendency in the dynamic analysis of multi-walled CNTs connected with the application of a multiple-elastic shell model which assumes that each of the concentric tubes of multi-wall carbon nanotubes is an individual elastic shell and coupled with adjacent tubes through Van der Waals interaction. The broader discussion of that class of problems was presented by Xu and Wang [97].

2.2   MOLECULAR SCALE METHODS

The modeling and simulation methods at molecular level usually employ atoms, molecules or their clusters as the basic units considered. The most popular methods include molecular mechanics (MM), MD and MC simulation. Modeling of polymer nano composites at this scale is predominantly directed toward the thermodynamics and kinetics of the formation, molecular structure and interactions. The diagram in Fig. 1.1 describes the equation of motion for each method and the typical properties predicted from each of them [98–102]. We introduce here the two widely used molecular scale methods: MD and MC.

2.2.1 MOLECULAR DYNAMICS

MD is a computer simulation technique that allows one to predict the time evolution of a system of interacting particles (e.g., atoms, molecules, granules, etc.) and estimate the relevant physical properties [103, 104]. Specifically, it generates such information as atomic positions, velocities and forces from which the macroscopic properties (e.g., pressure, energy, heat capacities) can be derived by means of statistical mechanics. MD simulation usually consists of three constituents: (i) a set of initial conditions (e.g., initial positions and velocities of all particles in the system); (ii) the interaction potentials to represent the forces among all the particles; (iii) the evolution of the system in time by solving a set of classical Newtonian equations of motion for all particles in the system. The equation of motion is generally given by

$$\vec{F}_i(t) = m_i \frac{d^2 \vec{r}_i}{dt^2} \tag{1}$$

where $\vec{F}_i(t)$ is the force acting on the i-th atom or particle at time t, which is obtained as the negative gradient of the interaction potential U, m_i is the atomic mass and $\vec{r}_i$ the atomic position. A physical simulation involves the proper selection of interaction potentials, numerical integration, periodic boundary conditions, and the controls of pressure and temperature to mimic physically meaningful thermodynamic ensembles. The interaction potentials together with their parameters, that is, the so-called force field, describe in detail how the particles in a system interact with each other, that is, how the potential energy of a system depends on the particle coordinates. Such a force field may be obtained by quantum method (e.g., AB initio), empirical method (e.g., LJ, Mores, and Born-Mayer) or quantum-empirical method (e.g., embedded atom model, glue model, bond order potential). The criteria for selecting a force field include the accuracy, transferability and computational speed. A typical interaction potential U may consist of a number of bonded and non-bonded interaction terms:

$$U\left(\vec{r}_1, \vec{r}_2, \vec{r}_3, \ldots, \vec{r}_n\right) = \sum_{i_{bond}}^{N_{bond}} U_{bond}\left(i_{bond}, \vec{r}_a, \vec{r}_b\right) + \sum_{i_{angle}}^{N_{angle}} U_{angle}\left(i_{angle}, \vec{r}_a, \vec{r}_b, \vec{r}_c\right)$$

$$+ \sum_{i_{torsion}}^{N_{torsion}} U_{torsion}\left(i_{torsion}, \vec{r}_a, \vec{r}_b, \vec{r}_c, \vec{r}_d\right)$$

$$+ \sum_{i_{inversion}}^{N_{inversion}} U_{inversion}\left(i_{inversion}, \vec{r}_a, \vec{r}_b, \vec{r}_c, \vec{r}_d\right)$$

$$+ \sum_{i=1}^{N-1} \sum_{j>1}^{N} U_{vdw}(i, j, \vec{r}_a, \vec{r}_b)$$

$$\sum_{i=1}^{N-1} \sum_{j>i}^{N} U_{electrostatic(i,j,\vec{r}_a,\vec{r}_b)} \tag{2}$$

The first four terms represent bonded interactions, that is, bond stretching *Ubond*, bond-angle bend *Uangle* and dihedral angle torsion *Utorsion* and inversion interaction *Uinversion*, while the last two terms are non-bonded interactions, that is, van der Waals energy *UvdW* and electrostatic energy *Uelectrostatic*. In the equation, $\vec{r}_a, \vec{r}_b, \vec{r}_c, \vec{r}_d$ are the positions of the atoms or particles specifically involved in a given interaction; *Nbond, Nangles, Ntorsion* and *Ninversion* stand for the total numbers of these respective interactions in the simulated system; *ibond, iangle, itorsion* and *iinversion* uniquely specify an individual interaction of each type; *i* and *j* in the van der Waals and electrostatic terms indicate the atoms involved in the interaction. There are many algorithms for integrating the equation of motion using finite difference methods. The algorithms of varlet, velocity varlct, lcap-frog and Beeman, are commonly used in MD simulations [104]. All algorithms assume that the atomic position $\vec{r}$, velocities $\vec{v}$ and accelerations $\vec{a}$, can be approximated by a Taylor series expansion:

$$\vec{r}(t+\delta t) = \vec{r}(t) + \vec{v}(t)\delta t + \frac{1}{2}\vec{a}(t)\delta^2 t + \dots \tag{3}$$

$$\vec{v}(t+\delta t) = \vec{v}(t) + \vec{a}(t)\delta t + \frac{1}{2}\vec{b}(t)\delta^2 t + \dots \tag{4}$$

$$\vec{a}(t+\delta t) = \vec{a}(t) + \vec{b}(t)\delta t + \dots \tag{5}$$

Generally speaking, a good integration algorithm should conserve the total energy and momentum and be time-reversible. It should also be easy to implement and computationally efficient, andpermit a relatively long time step. The Verlet algorithm is probably the most widely used method. It uses the positions $\overline{r(t)}$ and accelerations $\vec{a}(t)$ at time *t*, and the positions

$\vec{r}(t - \delta t)$ from the previous step (t–δ) to calculate the new positionsat (t+δt), we have:

$$\vec{r}(t+\delta t) = \vec{r}(t) + \vec{v}(t)\delta t + \frac{1}{2}\vec{a}(t)\delta t^2 + ... \tag{6}$$

$$\vec{r}(t-\delta t) = \vec{r}(t) - \vec{v}(t)\delta t + \frac{1}{2}\vec{a}(t)\delta t^2 + ... \tag{7}$$

$$\vec{r}(t+\delta t) = 2\vec{r}(t) - \vec{r}(t-\delta t) + \vec{a}(t)\delta t^2 + ... \tag{8}$$

The velocities at time t and $t + \dfrac{1}{2\delta t}$ can be estimated, respectively

$$\vec{v}(t) = \frac{\left[\vec{r}(t+\delta t) - \vec{r}(t-\delta t)\right]}{2\delta t} \tag{9}$$

$$\vec{v}\left(t+\frac{1}{2\delta t}\right) = \frac{\left[\vec{r}(t+\delta t) - \vec{r}(t-\delta t)\right]}{\delta t} \tag{10}$$

MD simulations can be performed in many different ensembles, such as grand canonical (μVT), micro canonical (NVE), canonical (NVT) and iso-thermal–isobaric (NPT). The constant temperature and pressure can be controlled by adding an appropriate thermostat (e.g., Berendsen, Nose, Nose-Hoover and Nose-Poincare) and barostat (e.g., Andersen, Hoover and Berendsen), respectively. Applying MD into polymer composites al-lows us to investigate into the effects of fillers on polymer structure and dynamics in the vicinity of polymer-filler interface and also to probe the effects of polymer-filler interactions on the materials properties.

2.2.2. MONTE CARLO

MC technique, also called Metropolis method [104], is a stochastic method that uses random numbers to generate a sample population of the system from which one can calculate the properties of interest. A MC simulation usually consists of three typical steps. In the first step, the physical prob-lem under investigation is translated into an analogous probabilistic or statistical model. In the second step, the probabilistic model is solved by a numerical stochastic sampling experiment. In the third step, the obtained data are analyzed by using statistical methods. MC provides only the information on equilibrium properties (e.g., free energy, phase equilibrium), different from MD,which gives non-equilibrium as well as equilibrium properties. In a NVT ensemble with N atoms, one hypothesizes a new configuration by arbitrarily or systematically moving one atom from position i→j. Due to such atomic movement, one can compute the change in the system Hamiltonian ΔH:

$$\Delta H = H(j)-H(i) \tag{11}$$

whereH(i) and H(j) are the Hamiltonian associated with the original and new configuration, respectively.

This new configuration is then evaluated according to the following rules. If ΔH<0 then the atomic movement would brings the system to a state of lower energy. Hence, the movement is immediately accepted and the displaced atom remains in its new position. If ΔH≥0, the move is accepted only with a certain probability Pi →j which is given by

$$Pi \rightarrow j \propto \exp\left(-\frac{\Delta H}{K_B T}\right) \tag{12}$$

where K_s is the Boltzmann constant. According to Metropolis et al. [105], one can generate a random number ζ between 0 and 1 and determine the new configuration according to the following rule:

$$æ \leq \exp\left(-\frac{\Delta H}{K_B T}\right); \text{The move is accepted;} \tag{13}$$

$$\zeta > \exp\left(-\frac{\Delta H}{K_B T}\right); \text{The move is not accepted.} \tag{14}$$

If the new configuration is rejected, one counts the original position as a new one and repeats the process by using other arbitrarily chosen atoms. In a μVT ensemble, one hypothesizes a new configuration j by arbitrarily choosing one atom and proposing that it can be exchanged by an atom of a different kind. This procedure affects the chemical composition of the system. Also, the move is accepted with a certain probability. However, one computes the energy change ΔU associated with the change in composition. The new configuration is examined according to the following rules. If $\Delta U < 0$ the movements of compositional change is accepted. However, if $\Delta U \geq 0$, the move is accepted with a certain probability which is given by

$$Pi \rightarrow \propto \exp\left(-\frac{\Delta H}{K_B T}\right) \tag{15}$$

where ΔU is the change in the sum of the mixing energy and the chemical potential of the mixture. If the new configuration is rejected one counts the original configuration as a new one and repeats the process by using some other arbitrarily or systematically chosen atoms. In polymer nano composites, MC methods have been used to investigate the molecular structure at nanoparticle surface and evaluate the effects of various factors.

2.3   MICRO SCALE METHODS

The modeling and simulation at micro scale aim to bridge molecular methods and continuum methods and avoid their shortcomings. Specifically, in nanoparticle-polymer systems, the study of structural evolution (i.e., dynamics of phase separation) involves the description of bulk flow (i.e., hydrodynamic behavior) and the interactions between nanoparticle and polymer components. Note that hydrodynamic behavior is relatively straight forward to handle by continuum methods but is very difficult and expensive to treat by atomistic methods. In contrast, the interactions between components can be examined at an atomistic level but are usually not straightforward to incorporate at the continuum level. Therefore, various simulation methods have been evaluated and extended to study the microscopic structure and phase separation of these polymer nano composites, including BD, DPD, LB, time-dependent Ginsburg-Landau (TDGL) theory, and dynamic DFT. In these methods, a polymer system is usually treated with a field description or microscopic particles that incorporate molecular details implicitly. Therefore, they are able to simulate the phenomena on length and time scales currently inaccessible by the classical MD methods.

2.3.1 BROWNIAN DYNAMICS

BD simulation is similar to MD simulations [105]. However; it introduces a few new approximations that allow one to perform simulations on the microsecond timescale whereas MD simulation is known up to a few nanoseconds. In BD the explicit description of solvent molecules used in MD is replaced with an implicit continuum solvent description. Besides, the internal motions of molecules are typically ignored, allowing a much larger time step than that of MD. Therefore, BD is particularly useful for systems where there is a large gap of time scale governing the motion of different components. For example, in polymer-solvent mixture, a short time-step is required to resolve the fast motion of the solvent molecules, whereas the evolution of the slower modes of the system requires a larger time step. However, if the detailed motion of the solvent molecules is concerned, they may be removed from the simulation and their effects on the polymer are represented by dissipative $(-\gamma P)$ and random $(\sigma\,\zeta(t))$ force terms. Thus, the forces in the governing Eq. (16) is replaced by a Langevin equation,

$$F_i(t) = \sum_{i \neq j} F_{ij}^c - \gamma P_i + \sigma \tau_i(t) \tag{16}$$

where F_{ij}^c is the conservative force of particle j acting on particle i, γ and σ are constants depending on the system, P_i the momentum of particle i, and $\zeta(t)$ a Gaussian random noise term. One consequence of this approximation of the fast degrees of freedom by fluctuating forces is that the energy and momentum are no longer conserved, which implies that the macroscopic behavior of the system will not be hydrodynamic. In addition, the effect of one solute molecule on another through the flow of solvent molecules is neglected. Thus, BD can only reproduce the diffusion properties but not the hydrodynamic flow properties since the simulation does not obey the Navier-Stokes equations.

2.3.2 DISSIPATIVE PARTICLE DYNAMICS

DPD was originally developed by Hoogerbrugge and Koelman [106]. It can simulate both Newtonian and non-Newtonian fluids, including polymer melts and blends, on microscopic length and time scales. Like MD and BD, DPD is a particle-based method. However, its basic unit is not a single atom or molecule but a molecular assembly (i.e., a particle). DPD particles are defined by their mass Mi, position ri and momentum Pi. The interaction force between two DPD particles i and j can be described by a sum of conservative F_{ij}^c, dissipative F_{ij}^D and random forces F_{ij}^R [107–109]:

$$F_{ij} = F_{ij}^C + F_{ij}^D + F_{ij}^R \qquad (17)$$

While the interaction potentials in MD are high-order polynomials of the distance between two particles, in DPD the potentials are softened so as to approximate the effective potential at microscopic length scales. The form of the conservative force in particular is chosen to decrease linearly with increasing r_{ij}. Beyond a certain cut-off separation r_c, the weight functions and thus the forces are all zero. Because the forces are pair wise and momentum is conserved, the macroscopic behavior directly incorporates Navier-Stokes hydrodynamics. However, energy is not conserved because of the presence of the dissipative and random force terms, which are similar to those of BD, but incorporate the effects of Brownian motion on larger length scales. DPD has several advantages over MD, for example, the hydrodynamic behavior is observed with far fewer particles than required in a MD simulation because of its larger particle size. Besides, its force forms allow larger time steps to be taken than those in MD.

2.3.3 LATTICE BOLTZMANN

LB [110] is another micro scale method that is suited for the efficient treatment of polymer solution dynamics. It has recently been used to investigate the phase separation of binary fluids in the presence of solid particles. The LB method is originated from lattice gas automaton, which is constructed as a simplified, fictitious molecular dynamic in which space, time and particle velocities are all discrete. A typical lattice gas automaton consists of a regular lattice with particles residing on the nodes. The main feature of the LB method is to replace the particle occupation variables (Boolean variables), by single-particle distribution functions (real variables) and neglect individual particle motion and particle-particle correlations in the kinetic equation. There are several ways to obtain the LB equation from either the discrete velocity model or the Boltzmann kinetic equation, and to derive the macroscopic Navier-Stokes equations from the LB equation. An important advantage of the LB method is that microscopic physical interactions of the fluid particles can be conveniently incorporated into the numerical model. Compared with the Navier-Stokes equations, the LB method can handle the interactions among fluid particles and reproduce the micro scale mechanism of hydrodynamic behavior. Therefore it belongs to the MD in nature and bridges the gap between the molecular level and macroscopic level. However, its main disadvantage is that it is typically not guaranteed to be numerically stable and may lead to physically unreasonable results, for instance, in the case of high forcing rate or high inter particle interaction strength.

2.3.4 TIME-DEPENDENT GINZBURG-LANDAU METHOD

TDGL is a micro scale method for simulating the structural evolution of phase-separation in polymer blends and block copolymers. It is based on the Cahn-Hilliard-Cook (CHC)non-linear diffusion equation for a binary blend and falls under the more general phase-field and reaction-diffusion models [111, 112]. In the TDGL method, a free-energy function is minimized to simulate a temperature quench from the miscible region of the phase diagram to the immiscible region. Thus, the resulting time-dependent structural evolution of the polymer blend can be investigated by solving the TDGL/CHC equation for the time dependence of the local blend concentration. Glotzer et al.[113] have discussed and applied this method to polymer blends and particle-filled polymer systems. This model reproduces the growth kinetics of the TDGL model, demonstrating that such quantities are insensitive to the precise form of the double-well potential of the bulk free-energy term. The TDGL and CDM methods have recently been used to investigate the phase-separation of polymer nanocomposite and polymer blends in the presence of nanoparticles [114–116].

2.3.5 DYNAMIC DFT METHOD

Dynamic DFT method is usually used to model the dynamic behavior of polymer systems and has been implemented in the software package Mesodyn™ from Accelrys [117]. The DFT models the behavior of polymer fluids by combining Gaussian mean-field statistics with a TDGL model for the time evolution of conserved order parameters. However, in contrast to traditional phenomenological free-energy expansion methods employed in the TDGL approach, the free energy is not truncated at a certain level, and instead retains the full polymer path integral numerically.

At the expense of a more challenging computation, this allows detailed information about a specific polymer system beyond simply the Flory-Huggins parameter and mobilities to be included in the simulation. In addition, viscoelasticity, which is not included in TDGL approaches, is included at the level of the Gaussian chains. A similar DFT approach has been developed by Doi et al. [118, 119] and forms the basis for their new software tool Simulation Utilities for Soft and Hard Interfaces (SUSHI), one of a suite of molecular and mesoscale modeling tools (called OCTA) developed for the simulation of polymer materials [119]. The essence of dynamic DFT method is that the instantaneous unique conformation distribution can be obtained from the off-equilibrium density profile by coupling a fictitious external potential to the Hamiltonian. Once such distribution is known, the free energy is then calculated by standard statistical thermodynamics. The driving force for diffusion is obtained from the spatial gradient of the first functional derivative of the free energy with respect to the density. Here, we describe briefly the equations for both polymer and particle in the diblock polymer-particle composites [120].

2.4   MESOSCALE AND MACROSCALE METHODS

Despite the importance of understanding the molecular structure and nature of materials, their behavior can be homogenized with respect to different aspects, which can be at different scales. Typically, the observed macroscopic behavior is usually explained by ignoring the discrete atomic and molecular structure and assuming that the material is continuously distributed throughout its volume. The continuum material is thus assumed to have an average density and can be subjected to body forces such as gravity and surface forces. Generally speaking, the macroscale methods (or called continuum methods hereafter) obey the fundamental laws of: (i) continuity, derived from the conservation of mass; (ii) equilibrium, derived from momentum considerations and Newton's second law; (iii) the moment of momentum principle, based on the model that the time rate of change of angular momentum with respect to an arbitrary point is equal to the resultant moment;(iv) conservation of energy, based on the first law of thermodynamics; and (v) conservation of entropy, based on the second law of thermodynamics. These laws provide the basis for the continuum model and must be coupled with the appropriate constitutive equations and the equations of state to provide all the equations necessary for solving a continuum problem. The continuum method relates the deformation of a continuous medium to the external forces acting on the medium and the resulting internal stress and strain. Computational approaches range from simple closed-form analytical expressions to micromechanics and complex structural mechanics calculations based on beam and shell theory. In this section, we introduce some continuum methods that have been used in polymer nano composites, including micromechanics models (e.g., Halpin-Tsai model, Mori-Tanaka model), equivalent-continuum model, self-consistent model and finite element analysis.

2.5   MICROMECHANICS

Since the assumption of uniformity in continuum mechanics may not hold at the micro scale level, micromechanics methods are used to express the continuum quantities associated with an infinitesimal material element in terms of structure and properties of the micro constituents. Thus, a central theme of micromechanics models is the development of a RVE to statistically represent the local continuum properties. The RVE is constructed to ensure that the length scale is consistent with the smallest constituent that has a first-order effect on the macroscopic behavior. The RVE is then used in a repeating or periodic nature in the full-scale model. The micromechanics method can account for interfaces between constituents, discontinuities, and coupled mechanical and non-mechanical properties. Our purpose is to review the micromechanics methods used for polymer nano composites. Thus, we only discuss here some important concepts of micromechanics as well as the Halpin-Tsai model and Mori-Tanaka model.

2.5.1   BASIC CONCEPTS

When applied to particle reinforced polymer composites, micromechanics models usually follow such basic assumptions as (i) linear elasticity of fillers and polymer matrix; (ii) the fillers are axisymmetric, identical in shape and size, and can be characterized by parameters such as aspect ratio; (iii) well-bonded filler-polymer interface and the ignorance of interfacial slip, filler-polymer de bonding or matrix cracking. The first concept is the linear elasticity, that is, the linear relationship between the total stress and infinitesimal strain tensors for the filler and matrix as expressed by the following constitutive equations:

$$\text{For filler} \quad \sigma^f = C^f \varepsilon^f \tag{18}$$

$$\text{For matrix} \quad \sigma^m = C^m \varepsilon^m \tag{19}$$

where C is the stiffness tensor. The second concept is the average stress and strain. Since the point wise stress field $\sigma(x)$ and the corresponding strain field $\varepsilon(x)$ are usually non-uniform in polymer composites, the volume-average stress σ and strain ε are then defined over the representative averaging volume V, respectively,

$$\bar{\sigma} = \frac{1}{v}\int \sigma(x)\,dv \tag{20}$$

$$\bar{\varepsilon} = \frac{1}{v}\int \tau(x)\,dv \tag{21}$$

Therefore, the average filler and matrix stresses are the averages over the corresponding volumes vf and vm, respectively,

$$\overline{\sigma_f} = \frac{1}{V_f}\int \sigma(x)\,dv \tag{22}$$

$$\overline{\sigma_m} = \frac{1}{V_f}\int \sigma(x)\,dv \tag{23}$$

The average strains for the fillers and matrix are defined, respectively, as

$$\overline{\varepsilon_f} = \frac{1}{V_f}\int \tau(x)\,dv \tag{24}$$

$$\overline{\varepsilon_m} = \frac{1}{V_m}\int \tau(x)\,dv \tag{25}$$

Based on the above definitions, the relationships between the filler and matrix averages and the overall averages can be derived as follows:

$$\bar{\sigma} = \overline{\sigma_f} v_f + \overline{\sigma_m} v_m \tag{26}$$

$$\bar{\varepsilon} = \overline{\varepsilon_f} v_f + \overline{\varepsilon_m} v_m \tag{27}$$

where vf and vm are the volume fractions of the fillers and matrix, respectively.

The third concept is the average properties of composites, which are actually the main goal of a micromechanics model. The average stiffness of the composite is the tensor C that maps the uniform strain to the average stress

$$\bar{\sigma} = \bar{\varepsilon} C \tag{28}$$

The average compliance S is defined in the same way:

$$\bar{\varepsilon} = \bar{\sigma} S \tag{29}$$

Another important concept is the strain-concentration and stress-concentration tensors A and B, which are basically the ratios between the average filler strain (and stress) and the corresponding average of the composites.

$$\overline{\varepsilon}_f = \overline{\overline{\varepsilon}} A \tag{30}$$

$$\overline{\sigma}_f = \overline{\overline{\sigma}} B \tag{31}$$

Using the above concepts and equations, the average composite stiffness can be obtained from the strain concentration tensor A and the filler and matrix properties:

$$C = C_m + v_f \left[(C)_f - C_m \right)_A \tag{32}$$

2.5.2   HALPIN-TSAI MODEL

The Halpin-Tsai model is a well-known composite theory to predict the stiffness of unidirectional composites as a functional of aspect ratio. In this model, the longitudinal E_{11} and transverse E_{22} engineering moduli are expressed in the following general form:

$$\frac{E}{E_m} = \frac{1+\zeta\eta v_f}{1-\eta v_f} \tag{33}$$

where E and E_m represent the Young's modulus of the composite and matrix, respectively, is the volume fraction of filler, and η is given by:

$$\eta = \frac{\dfrac{E}{E_m}-1}{\dfrac{E_f}{E_m}+\zeta_f} \tag{34}$$

where E_f represents the Young's modulus of the filler and ζ_f the shape parameter depending on the filler geometry and loading direction. When calculating longitudinal modulus E_{11}, ζ_f is equal to l/t, and when calculating transverse modulus E_{22}, ζ_f is equal to w/t. Here, the parameters of l, w and t are the length, width and thickness of the dispersed fillers, respectively. If $\zeta_f \rightarrow 0$, the Halpin-Tsai theory converges to the inverse rule of mixture (lower bound):

$$\frac{1}{E} = \frac{v_f}{E_f} + \frac{1-v_f}{E_m} \tag{35}$$

Conversely, if $\zeta_f \rightarrow \infty$, the theory reduces to the rule of mixtures (upper bound),

$$E = E_f v_f + E_m \left(1 - v_f\right) \tag{36}$$

2.5.3 MORI–TANAKA MODEL

The Mori-Tanaka model is derived based on the principles of Eshelby's inclusion model for predicting an elastic stress field in and around ellipsoidal filler in an infinite matrix. The complete analytical solutions for longitudinal E_{11} and transverse E_{22} elastic moduli of an isotropic matrix filled with aligned spherical inclusion are [121]:

$$\frac{E_{11}}{E_m} = \frac{2A_0}{A_0 + v_f\left(A_1 + 2v_0 A_2\right)} \tag{37}$$

$$\frac{E_{22}}{E_m} = \frac{2A_0}{2A_0 + v_f \left(-2A_3 + \left(1 - v_0 A_4\right) + \left(1 + v_0\right) A_5 A_0\right)} \tag{38}$$

where *Em* represents the Young's modulus of the matrix, *vf* the volume fraction of filler, v_0 the Poisson's ratio of the matrix, parameters, A_0, A_1,... , A_5 are functions of the Eshelby's tensor and the properties of the filler and the matrix, including Young's modulus, Poisson's ratio, filler concentration and filler aspect ratio [121].

2.5.4 EQUIVALENT-CONTINUUM AND SELF-SIMILAR

Numerous micromechanical models have been successfully used to predict the macroscopic behavior of fiber-reinforced composites. However, the direct use of these models for nanotubes-reinforced composites is doubtful due to the significant scale difference between nanotubes and typical carbon fiber. Recently, two methods have been proposed for modeling the mechanical behavior of single walled carbon nanotubes (SWCN) composites:

Equivalent-continuum approach and self-similar approaches [122].The equivalent-continuum approach was proposed by Odegard et al. [123]. In this approach, MD was used to model the molecular interactions between SWCN-polymer and a homogeneous equivalent-continuum reinforcing element (e.g., a SWCN surrounded polymer) was constructed. Then, micromechanics are used to determine the effective bulk properties of the equivalent-continuum reinforcing element embedded in a continuous polymer. The equivalent-continuum approach consists of four major steps, as briefly described below.

Step 1: MD simulation is used to generate the equilibrium structure of a SWCN-polymer composite and then to establish the RVE of the molecular model and the equivalent-continuum model.

Step 2: The potential energies of deformation for the molecular model and effective fiber are derived and equated for identical loading conditions. The bonded and non-bonded interactions within a polymer molecule are quantitatively described by MM. For the SWCN/polymer system, the total potential energy U^m of the molecular model is,

$$U^m = \sum U^r \left(K_r\right) + \sum U^\theta \left(k_\theta\right) + \sum U^{vdw} \left(k_{vdw}\right) \tag{39}$$

where U^r, U^θ and U^{vdw} are the energies associated with covalent bond stretching, bond-angle bending, andvan der Waals interactions, respectively. An equivalent-truss model of the RVE is used as an intermediate step to link the molecular and equivalent-continuum models. Each atom in the molecular model is represented by a pin-joint, and each truss element represents an atomic bonded or non-bonded interaction. The potential energy of the truss model is

$$U^t = \sum U^a \left(E^a \right) + \sum U^b \left(E^b \right) + \sum U^c \left(E^c \right) \tag{40}$$

where U^a, U^b and U^c are the energies associated with truss elements that represent covalent bond stretching, bond-angle bending, andvan der Waals interactions, respectively. The energies of each truss element are a function of the Young's modulus, E.

Step 3: A constitutive equation for the effective fiber is established. Since the values of the elastic stiffness tensor components are not known a priori, a set of loading conditions are chosen such that each component is uniquely determined from

$$U^f = U^t = U^m \tag{41}$$

Step 4: Overall constitutive properties of the dilute and unidirectional SWCN/polymer composite are determined with Mori-Tanaka model with the mechanical properties of the effective fiber and the bulk polymer. The layer of polymer molecules that are near the polymer/nanotubes interface is included in the effective fiber, and it is assumed that the matrix polymer surrounding the effective fiber has mechanical properties equal to those of the bulk polymer. The self-similar approach was proposed by Pipes and Hubert [124],which consists of three major steps.

First, a helical array of SWCNs is assembled. This array is termed as the SWCN nano array where 91 SWCNs make up the cross-section of the helical nano array. Second, the SWCN nano arrays is surrounded by a polymer matrix and assembled into a second twisted array, termed as the SWCN nano wire. Third, the SWCN nano wires are further impregnated with a polymer matrix and assembled into the final helical array the SWCN microfiber. The self-similar geometries described in the nano array, nano wire and microfiber allow the use of the same mathematical and geometric model for all three geometries [124].

2.5.5 FINITE ELEMENT METHOD

FEM is a general numerical method for obtaining approximate solutions in space to initial-value and boundary-value problems including time-dependent processes. It employs preprocessed mesh generation, which enables the model to fully capture the spatial discontinuities of highly inhomogeneous materials. It also allows complex, non-linear tensile relationships to be incorporated into the analysis. Thus, it has been widely used in mechanical, biological and geological systems. In FEM, the entire domain of interest is spatially discretized into an assembly of simply shaped sub domains (e.g., hexahedra or tetrahedral in three dimensions, and rectangles or triangles in two dimensions) without gaps and without overlaps. The sub domains are interconnected at joints (i.e., nodes). The energy in FEM is taken from the theory of linear elasticity and thus the input parameters are simply the elastic moduli and the density of the material. Since these parameters are in agreement with the values computed by MD, the simulation is consistent across the scales. More specifically, the total elastic energy in the absence of tractions and body forces within the continuum model is given by [125].

$$U^f = U^t = U^m \tag{42}$$

$$U_k = \frac{1}{2}\int dr p(r) \left| U_r \right|^2 \tag{43}$$

$$U_v = \frac{1}{2}\int dr \sum_{\mu\nu\sigma\lambda=1}^{a} \varepsilon_{\mu\nu}(r) C_{\mu\nu\lambda\sigma^3\lambda\sigma(r)} \tag{44}$$

where U_v is the Hookian potential energy term, which is quadratic in the symmetric strain tensor e, contracted with the elastic constant tensor C. The Greek indices (i.e., m, n, l, s) denote Cartesian directions. The kinetic energy U_k involves the time rate of change of the displacement field $\dot{U}$, and the mass density ρ.

These are fields defined throughout space in the continuum theory. Thus, the total energy of the system is an integral of these quantities over the volume of the sample dv. The FEM has been incorporated in some commercial software packages and open source codes (e.g., ABAQUS, ANSYS, Palmyra and OOF) and widely used to evaluate the mechanical properties of polymer composites. Some attempts have recently been made to apply the FEM to nanoparticle-reinforced polymer nano composites. In order to capture the multi scale material behaviors, efforts are also underway to combine the multi scale models spanning from molecular to macroscopic levels [126, 127].

2.6   MULTI SCALE MODELING OF MECHANICAL PROPERTIES

In Odegard's study [123], a method has been presented for linking atomistic simulations of nano-structured materials to continuum models of the corresponding bulk material. For a polymer composite system reinforced with single walled carbon nanotubes (SWNT), the method provides the steps whereby the nanotubes, the local polymer near the nanotubes, and the nanotubes/polymer interface can be modeled as an effective continuum fiber by using an equivalent-continuum model. The effective fiber retains the local molecular structure and bonding information, as defined by molecular dynamics, and serves as a means for linking the equivalent-continuum and micromechanics models. The micromechanics method is then available for the prediction of bulk mechanical properties of SWNT/polymer composites as a function of nanotubes size, orientation, and volume fraction. The utility of this method was examined by modeling tow composites that both having an interface. The elastic stiffness constants of the composites were determined for both aligned and three-dimensional randomly oriented nanotubes, as a function of nanotubes length and volume fraction. They used Mori-Tanaka model [128] for random and oriented fibers position and compare their model with mechanical properties, the interface between fiber and matrix was assumed perfect. Motivated by micrographs showing that embedded nanotubes often exhibit significant curvature within the polymer, Fisher et al. [129] have developed a model combining finite element results and micromechanical methods (Mori-Tanaka) to determine the effective reinforcing modulus (ERM) of a wavy embedded nanotubes with perfect bonding and random fiber orientation assumption. This ERM is then used within a multiphase micromechanics model to predict the effective modulus of a polymer reinforced with a distribution of wavy nanotubes. We found that even slight nanotubes curvature significantly reduces the effective reinforcement when compared to straight nanotubes. These results suggest that nanotubes waviness may be an additional mechanism limiting the modulus enhancement of nanotubes-reinforced polymers. Bradshaw et al. [130] investigated the degree to which the characteristic waviness of nanotubes embedded in polymers can impact the effective stiffness of these materials. A 3D finite element model of a single infinitely long sinusoidal fiber within an infinite matrix is used to numerically compute the dilute strain concentration tensor. A Mori-Tanaka model uses this tensor to predict the effective modulus of the material with aligned or randomly oriented inclusions. This hybrid finite element micromechanical modeling technique is a powerful extension of general micromechanics modeling and can be applied to any composite microstructure containing non-ellipsoidal inclusions. The results demonstrate that nanotubes waviness results in a reduction of the effective modulus of the composite relative to straight nanotubes reinforcement. The degree of reduction is dependent on the ratio of the sinusoidal wavelength to the nanotubes diameter. As this wavelength ratio increases, the effective stiffness of a composite with randomly oriented wavy nanotubes converges to the result obtained with straight nanotubes inclusions.

The effective mechanical properties of carbon nanotubes-based composites are evaluated by Liu and Chen [131] using a 3D nanoscale RVE based on 3D elasticity theory and solved by the finite element method. Formulas to extract the material constants from solutions for the RVE under three loading cases are established using the elasticity. An extended rule of mixtures, which can be used to estimate the Young's modulus in the axial direction of the RVE and to validate the numerical solutions for short CNTs, is also derived using the strength of materials theory. Numerical examples using the FEM to evaluate the effective material constants of a CNT-based composites are presented, which demonstrate that the reinforcing capabilities of the CNTs in a matrix are significant. With only about 2% and 5% volume fractions of the CNTs in a matrix, the stiffness of the composite in the CNT axial direction can increase as many as 0.7 and 9.7 times for the cases of short and long CNT fibers, respectively. These simulation results, which are believed to be the first of its kind for CNT-based composites, are consistent with the experimental results reported in the literature Schadler et al. [75], Wagner et al. [70]; Qian et al. [66]. The developed extended rule of mixtures is also found to be quite effective in evaluating the stiffness of the CNT-based composites in the CNT axial direction. Many research issues need to be addressed in the modeling and simulations of CNTs in a matrix material for the development of nano composites. Analytical methods and simulation models to extract the mechanical properties of the CNT-based nano composites need to be further developed and verified with experimental results. The analytical method and simulation approach developed in this paper are only a preliminary study. Different type of RVEs, load cases and different solution methods should be investigated. Different interface conditions, other than perfect bonding, need to be investigated using different models to more accurately account for the interactions of the CNTs in a matrix material at the nanoscale. Nanoscale interface cracks can be analyzed using simulations to investigate the failure mechanism in nanomaterials. Interactions among a large number of CNTs in a matrix can be simulated if the computing power is available. Single-walled and multi-walled CNTs as reinforcing fibers in a matrix can be studied by simulations to find out their advantages and disadvantages. Finally, large multi scale simulation models for CNT-based composites, which can link the models at the nano, micro and macro scales, need to be developed, with the help of analytical and experimental work [131].The three RVEs proposed in Ref. [132] and shown in Fig. 2.1 are relatively simple regarding the models and scales and pictures in Fig. 2.2 are Three loading cases for the cylindrical RVE. However, this is only the first step toward more sophisticated and large-scale simulations of CNT-based composites. As the computing power and

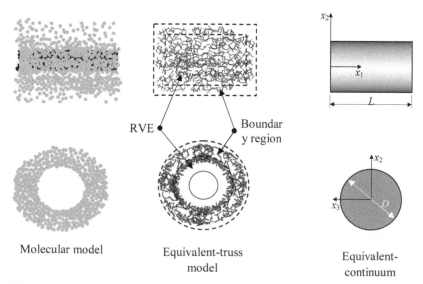

Molecular model   Equivalent-truss   Equivalent-
                              model   continuum

FIGURE 2.1   Equivalent-continuum modeling of effective's fiber.

confidence in simulations of CNT-based composites increase, large scale 3D models containing hundreds or even more CNTs, behaving linearly or non-linearly, with coatings or of different sizes, distributed evenly or randomly, can be employed to investigate the interactions among the CNTs in a matrix and to evaluate the effective material properties. Other numerical methods can also be attempted for the modeling and simulations of CNT-based composites, which may offer some advantages over the FEM approach. For example, the boundary element method, Liu et al. [132]; Chen and Liu [133], accelerated with the fast multi pole techniques, Fu et al. [134]; Nishimura et al. [135], and the mesh free methods (Qian et al. [136])may enable one to model an RVE with thousands of CNTs in a matrix on a desktop computer. Analysis of the CNT-based composites using the boundary element method is already underway and will be reported subsequently (Fig. 2.3).

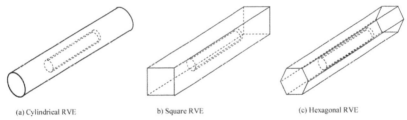

(a) Cylindrical RVE   b) Square RVE   (c) Hexagonal RVE

FIGURE 2.2   Three nanoscale representative volume elements for the analysis of CNT based nano composites.

The effective mechanical properties of CNT based composites are evaluated using square RVEs based on 3D elasticity theory and solved by the FEM. Formulas to extract the effective material constants from solutions for the square RVEs under two loading cases are established based on elasticity. Square RVEs with multiple CNTs are also investigated in evaluating the Young's modulus and Poisson's ratios in the transverse plane. Numerical examples using the FEM are presented, which demonstrate that the load-carrying capabilities of the CNTs in a matrix are significant. With the addition of only about 3.6% volume fraction of the CNTs in a matrix, the stiffness of the composite in the CNT axial direction can increase as much as 33% for the case of long CNT fibers [136]. These simulation results are consistent with both the experimental ones reported

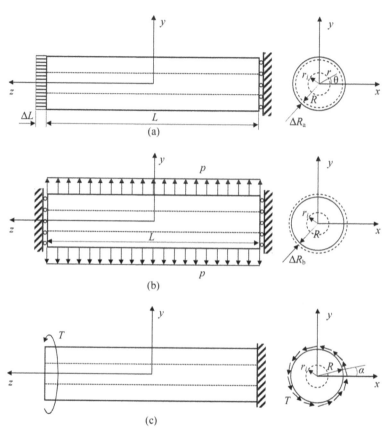

FIGURE 2.3 Three loading cases for the cylindrical RVE used to evaluate the effective material properties of the CNT-based composites. (a) Under axial stretch DL; (b) under lateral uniform load P; (c) under Torsional load T.

in the literature [66, 70, 75, 131, 137]. It is also found that cylindrical RVEs tend to overestimate the effective Young's moduli due to the fact that they overestimate the volume fractions of the CNTs in a matrix. The square RVEs, although more demanding in modeling and computing, may be the preferred model in future simulations for estimating the effective material constants, especially when multiple CNTs need to be considered. Finally, the rules of mixtures, for both long and short CNT cases, are found to be quite accurate in estimating the effective Young's moduli in the CNT axial direction. This may suggest that 3D FEM modeling may not be necessary in obtaining the effective material constants in the CNT direction, as in the studies of the conventional fiber reinforced composites. Efforts in comparing the results presented in this paper using the continuum approach directly with the MD simulations are underway. This is feasible now only for a smaller RVE of one CNT embedded in a matrix. In future research, the MD and continuum approach should be integrated in a multi scale modeling and simulation environment for analyzing the CNT-based composites. More efficient models of the CNTs in a matrix also need to be developed, so that a large number of CNTs, in different shapes and forms (curved or twisted), or randomly distributed in a matrix, can be modeled. The ultimate validation of the simulation results should be done with the nanoscale or micro scale experiments on the CNT reinforced composites [136].

Griebel and Hamaekers [45]reviewed the basic tools used in computational nano mechanics and materials, including the relevant underlying principles and concepts. These tools range from subatomic Ab initio methods to classical molecular dynamics and multiple-scale approaches. The energetic link between the quantum mechanical and classical systems has been discussed, and limitations of the standing alone molecular dynamics simulations have been shown on a series of illustrative examples. The need for multiscale simulation methods to take nanoscale aspects of material behavior was therefore emphasized; that was followed by a review and classification of the mainstream and emerging multiscale methods. These simulation methods include the broad areas of quantum mechanics, molecular dynamics and multiple-scale approaches, based on coupling the atomistic and continuum models. They summarize the strengths and limitations of currently available multiple-scale techniques, where the emphasis is made on the latest perspective approaches, such as the bridging scale method, multiscale boundary conditions, and multiscale fluidics. Example problems, in which multiple-scale simulation methods yield equivalent results to full atomistic simulations at fractions of the computational cost, were shown. They compare their results with Odegard, et al. [123]; the micro mechanic method was BEM Halpin-Tsai equation [107]with aligned fiber by perfect bonding.

The solutions of the strain-energy-changes due to a SWNT embedded in an infinite matrix with imperfect fiber bonding are obtained through numerical method by Wan, et al. [138]. A "critical" SWNT fiber length is defined for full load transfer between the SWNT and the matrix, through the evaluation of the strain-energy-changes for different fiber lengths

The strain-energy-change is also used to derive the effective longitudinal Young's modulus and effective bulk modulus of the composite, using a dilute solution. The main goal of their research was investigation of strain-energy-change due to inclusion of SWNT using FEM. To achieve full load transfer between the SWNT and the matrix, the length of SWNT fibers should be longer than a 'critical' length if no weak inter phase exists between the SWNT and the matrix [138].

A hybrid atomistic/continuum mechanics method is established in the Feng, et al. [139] study the deformation and fracture behaviors of carbon nanotubes (CNTs) in composites. The unit cell containing a CNT embedded in a matrix is divided in three regions, which are simulated by the atomic-potential method, the continuum method based on the modified Cauchy-Born rule, and the classical continuum mechanics, respectively. The effect of CNT interaction is taken into account via the Mori-Tanaka effective field method of micromechanics. This method not only can predict the formation of Stone-Wales (5–7–7–5) defects, but also simulate the subsequent deformation and fracture process of CNTs. It is found that the critical strain of defect nucleation in a CNT is sensitive to its chiral angle but not to its diameter. The critical strain of Stone-Wales defect formation of zigzag CNTs is nearly twice that of armchair CNTs. Due to the constraint effect of matrix, the CNTs embedded in a composite are easier to fracture in comparison with those not embedded. With the increase in the Young's modulus of the matrix, the critical breaking strain of CNTs decreases.

Estimation of effective elastic moduli of nano composites was performed by the version of effective field method developed in the framework of quasi-crystalline approximation when the spatial correlations of inclusion location take particular ellipsoidal forms [140]. The independent justified choice of shapes of inclusions and correlation holes provide the formulae of effective moduli, which are symmetric, completely explicit and easily to use. The parametric numerical analyzes revealed the most sensitive parameters influencing the effective moduli which are defined by the axial elastic moduli of nano fibers rather than their transversal moduli as well as by the justified choice of correlation holes, concentration and prescribed random orientation of nano fibers [140].

Li and Chou [141, 142] have reported a multi scale modeling of the compressive behavior of carbon nano tube/polymer composites. The nano tube is modeled at the atomistic scale, and the matrix deformation is analyzed by the continuum finite element method. The nano tube and polymer matrix are assumed to be bonded by van der Waals interactions at the interface. The stress distributions at the nano tube/polymer interface under isostrain and isostress loading conditions have been examined they have used beam elements for SWCNT using molecular structural mechanics, truss rod for van der Waals links and cubic elements for matrix the rule of mixture was used as for comparison in this research. The buckling forces of nano tube/polymer composites for different nano tube lengths and diameters are computed. The results indicate that continuous nanotubes can most effectively enhance the composite buckling resistance.

Anumandla and Gibson [143] describes an approximate, yet comprehensive, closed form micromechanics model for estimating the effective elastic modulus of carbon nano tube-reinforced composites. The model incorporates the typically observed nano tube curvature, the nano tube length, and both 1D and 3D random arrangement of the nanotubes. The analytical results obtained from the closed form micromechanics model for nanoscale RVEs and results from an equivalent finite element model for ERM of the nano tube reveal that the reinforcing modulus is strongly dependent on the waviness, wherein, even a slight change in the nano tube curvature can induce a prominent change in the effective reinforcement provided. The micromechanics model is also seen to produce reasonable agreement with experimental data for the effective tensile modulus of composites reinforced with multi-walled nanotubes (MWNTs) and having different MWNT volume fractions.

Effective elastic properties for carbon nano tube reinforced composites are obtained through a variety of micromechanics techniques [144]. Using the in-plane elastic properties of graphene, the effective properties of carbon nanotubes are calculated using a composite cylinders micromechanics technique as a first step in a two-step process. These effective properties are then used in the self-consistent and Mori-Tanaka methods to obtain effective elastic properties of composites consisting of aligned single or multi-walled carbon nanotubes embedded in a polymer matrix. Effective composite properties from these averaging methods are compared to a direct composite cylinders approach extended from the work of Hashin and Rosen [145] and Christensen and Lo [146]. Comparisons with finite element simulations are also performed. The effects of an inter phase layer between the nanotubes and the polymer matrix as result of functionalization is also investigated using a multilayer composite cylinders approach. Finally, the modeling of the clustering of nanotubes into bundles due to inter atomic forces is accomplished herein using a tessellation method in conjunction with a multiphase Mori-Tanaka technique. In addition to aligned nano tube composites, modeling of the effective elastic properties of randomly dispersed nanotubes into a matrix is performed using the Mori-Tanaka method, and comparisons with experimental data are made.

Selmi, et al. [147] deal with the prediction of the elastic properties of polymer composites reinforced with single walled carbon nanotubes. Their contribution is the investigation of several micromechanical models, while most of the papers on the subject deal with only one approach. They implemented four homogenization schemes, a sequential one and three others based on various extensions of the Mori-Tanaka (M-T) mean-field homogenization model: two-level (M-T/M-T), two-step (M-T/M-T) and two-step (M-T/Voigt). Several composite systems are studied, with various properties of the matrix and the graphene, short or long nanotubes, fully aligned or randomly oriented in 3D or 2D. Validation targets are experimental data or finite element results, either based on a 2D periodic unit cell or a 3D RVE. The comparative study showed that there are cases where all micromechanical models give adequate predictions, while for some composite materials and some properties, certain models fail in a rather spectacular fashion. It was found that the two-level (M-T/M-T) homogenization model gives the best predictions in most cases. After the characterization of the discrete nano tube structure using a homogenization method based on energy equivalence, the sequential, the two-step (M-T/M-T), the two-step (M-T/Voigt), the two-level (M-T/M-T) and finite element models were used to predict the elastic properties of SWNT/polymer composites. The data delivered by the micromechanical models are compared against those obtained by finite element analyzes or experiments. For fully aligned, long nano tube polymer composite, it is the sequential and the two-level (M-T/M-T) models, which delivered good predictions. For all composite morphologies (fully aligned, two-dimensional in-plane random orientation, and three-dimensional random orientation), it is the two-level (M-T/M-T) model,which gave good predictions compared to finite element and experimental results in most situations. There are cases where other micromechanical models failed in a spectacular way.

Luo, et al. [148] have used multiscale homogenization (MH) and FEM for wavy and straight SWCNTs, they have compare their results with Mori-Tanaka, Cox, Halpin-Tsai, Fu, et al. [149], Lauke [149], Tserpes, et al. [150] used 3D elastic beam for C-C bond and, 3D space frame for CNT and progressive fracture model for prediction of elastic modulus, they used rule of mixture for compression of their results. Their assumption was embedded a single SWCNT in polymer with Perfect bonding. The multiscale modeling, Monte Carlo, FEM and using equivalent continuum method was used by Spanos and Kontsos [151] and compared with Zhu, et al. [152] and Paiva, et al. [153]'s results.

Bhuiyan et al. [154] studied the effective modulus of CNT/PP composites using FEA of a 3D RVE which includes the PP matrix, multiple CNTs and CNT/PP inter phase and accounts for poor dispersion and non homogeneous distribution of CNTs within the polymer matrix, weak CNT/polymer interactions, CNT agglomerates of various sizes and CNTs orientation and waviness. Currently, there is no other model, theoretical or numerical, that accounts for all these experimentally observed phenomena and captures their individual and combined effect on the effective modulus of nano composites. The model is developed using input obtained from experiments and validated against experimental data. CNT reinforced PP composites manufactured by extrusion and injection molding are characterized in terms of tensile modulus, thickness and stiffness of CNT/PP inter phase, size of CNT agglomerates and CNT distribution using tensile testing, AFM and SEM, respectively. It is concluded that CNT agglomeration and waviness are the two dominant factors that hinder the great potential of CNTs as polymer reinforcement. The proposed model provides the upper and lower limit of the modulus of the CNT/PP composites and can be used to guide the manufacturing of composites with engineered properties for targeted applications. CNT agglomeration can be avoided by employing processing techniques such as sonication of CNTs, stirring, calendaring, etc., whereas CNT waviness can be eliminated by increasing the injection pressure during molding and mainly by using CNTs with smaller aspect ratio. Increased pressure during molding can also promote the alignment of CNTs along the applied load direction. The 3D modeling capability presented in this study gives an insight on the upper and lower bound of the CNT/PP composites modulus quantitatively by accurately capturing the effect of various processing parameters. It is observed that when all the experimentally observed factors are considered together in the FEA the modulus prediction is in good agreement with the modulus obtained from the experiment. Therefore, it can be concluded that the FEM models proposed in this study by systematically incorporating experimentally observed characteristics can be effectively used for the determination of mechanical properties of nanocomposite materials. Their result is in agreement with the results reported in Ref. [155]. The theoretical micromechanical models, shown in Fig. 2.4, are used to confirm that our FEM model predictions follow the same trend with the one predicted by the models as expected.

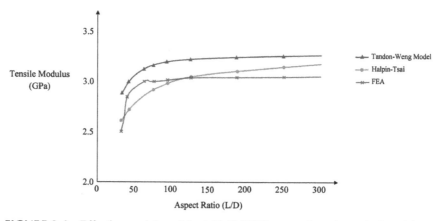

FIGURE 2.4   Effective modulus of 5 wt. % CNT/PP composites: theoretical models vs. FEA.

For reasons of simplicity and in order to minimize the mesh dependency on the results the hollow CNTs are considered as solid cylinders of circular cross-sectional area with an equivalent average diameter, shown in Fig. 2.5, calculated by equating the volume of the hollow CNT to the solid one [155].

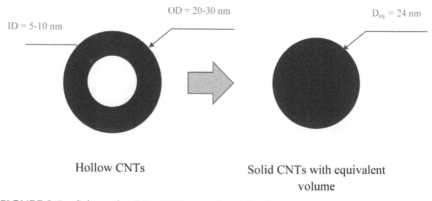

FIGURE 2.5   Schematic of the CNTs considered for the FEA.

The micromechanical models used for the comparison was Halpin-Tsai (H-T) [156] and Tandon-Weng (T-W) [121] model and the comparison was performed for 5 wt.% CNT/PP. It was noted that the H-T model results to lower modulus compared to FEA because H-T equation does not account for maximum packing fraction and the arrangement of the reinforcement in the composite. A modified H-T model that account for this has been proposed in the literature [157]. The effect of maximum packing fraction and the arrangement of the reinforcement within the composite become less significant at higher aspect ratios [158].

A finite element model of carbon nanotubes, interphase and its surrounding polymer is constructed to study the tensile behavior of embedded short carbon nanotubes in polymer matrix in presence of van der Waals interactions in interphase region by Shokrieh and Rafiee [159]. The interphase is modeled using non-linear spring elements capturing the force-distance curve of van der Waals interactions. The constructed model is subjected to tensile loading to extract longitudinal Young's modulus. The obtained results of this work have been compared with the results of previous research of the same authors [160] on long embedded carbon nanotubes in polymer matrix. It shows that the capped short carbon nanotubes reinforce polymer matrix less efficient than long CNTs.

Despite the fact that researches have succeeded to grow the length of CNTs up to 4 cm as a world record in US Department of Energy Los Alamos National Laboratory [161] and also there are some evidences on producing CNTs with lengths up to millimeters [162], CNTs are commercially available in different lengths ranging from 100 nm to approximately 30 lm in the market based on employed process of growth. Chemists at Rice University have identified a chemical process to cut CNTs into short segments. As a consequent, it can be concluded that the SW-CNTs with lengths smaller than 1000 nm do not contribute significantly in reinforcing polymer matrix. On the other hand, the efficient length of reinforcement for a CNT with (10, 10) index is about 1.2 lm and short CNT with length of 10.8 lm can play the same role as long CNT reflecting the uppermost value reported in our previous research. Finally, it is shown that the direct use of Halpin-Tsai equation to predict the modulus of SWCNT/composites overestimates the results. It is also observed that application of previously developed long equivalent fiber stiffness is a good candidate to be used in Halpin-Tsai equations instead of Young's modulus of CNT. Halpin-Tsai equation is not an appropriate model for smaller lengths, since there is not any reinforcement at all for very small lengths [159].

Earlier, a nano-mechanical model has been developed by Chowdhury et al. [163] to calculate the tensile modulus and the tensile strength of randomly oriented short carbon nanotubes (CNTs) reinforced nano composites, considering the statistical variations of diameter and length of the CNTs. According to this model, the entire composite is divided into several composite segments which contain CNTs of almost the same diameter and length. The tensile modulus and tensile strength of the composite are then calculated by the weighted sum of the corresponding modulus and strength of each composite segment. The existing micromechanical approach for modeling the short fiber composites is modified to account for the structure of the CNTs, to calculate the modulus and the strength of each segmented CNT reinforced composites. Multi-walled CNTs with and without intertube bridging (see Fig. 2.6) have been considered. Statistical variations of the diameter and length of the CNTs are modeled by a normal distribution. Simulation results show that CNTs intertube bridging; length and diameter affect the nano composites modulus and strength. Simulation results have been compared with the available experimental results and the comparison concludes that the developed model can be effectively used to predict tensile modulus and tensile strength of CNTs reinforced composites.

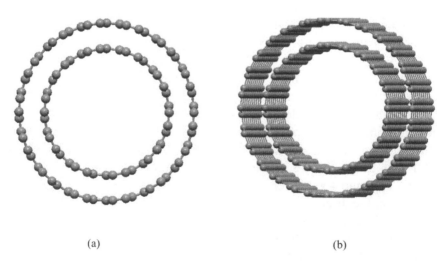

(a)   (b)

FIGURE 2.6   Schematic of MWNT with intertube bridging (a) Top view and (b) oblique view

The effective elastic properties of carbon nanotubes-reinforced polymers have been evaluated by Tserpes and Chanteli [104] as functions of material and geometrical parameters using a homogenized RVE. The RVE consists of the polymer matrix, a multi-walled carbon nanotubes (MWCNT) embedded into the matrix and the interface between them. The parameters considered are the nanotubes aspect ratio, the nanotubes volume fraction as well as the interface stiffness and thickness. For the MWCNT, both isotropic and orthotropic material properties have been considered. Analyzes have been performed by means of a 3D FE model of the RVE. The results indicate a significant effect of nanotubes volume fraction. The effect of nanotubes aspect ratio appears mainly at low values and diminishes after the value of 20. The interface mostly affects the effective elastic properties at the transverse direction. Having evaluated the effective elastic properties of the MWCNT-polymer at the microscale, the RVE has been used to predict the tensile modulus of a polystyrene specimen reinforced by randomly aligned MWCNTs for which experimental data exist in the literature. A very good agreement is obtained between the predicted and experimental tensile moduli of the specimen. The effect of nanotubes alignment on the specimen's tensile modulus has been also examined and found to be significant since as misalignment increases the effective tensile modulus decreases radically. The proposed model can be used for the virtual design and optimization of CNT-polymer composites since it has proven capable of assessing the effects of different material and geometrical parameters on the elastic properties of the composite and predicting the tensile modulus of CNT-reinforced polymer specimens.

2.7 MODELING OF THE INTERFACE

2.7.1 NANOTUBES-POLYMER INTERFACE IN COMPOSITE APPLICATION

The superior mechanical properties of the nanotubes alone do not ensure mechanically superior composites because the composite properties are strongly influenced by the mechanics that govern the nanotubes-polymer interface. Typically in composites, the constituents do not dissolve or merge completely and therefore, normally, exhibit an interface between one another, which can be considered as a different material with different mechanical properties. The structural strength characteristics of composites greatly depend on the nature of bonding at the interface, the mechanical load transfer from the matrix (polymer) to the nanotubes and the yielding of the interface. As an example, if the composite is subjected to tensile loading and there exists perfect bonding between the nanotubes and polymer and/or a strong interface then the load (stress) is transferred to the nanotubes; since the tensile strength of the nanotubes (or the interface) is very high the composite can withstand high loads. However, if the interface is weak or the bonding is poor, on application of high loading either the interface fails or the load is not transferred to the nanotubes and the polymer fails due to their lower tensile strengths. Consider another example of transverse crack propagation. When the crack reaches the interface, it will tend to propagate along the interface, since the interface is relatively weaker (generally) than the nanotubes (with respect to resistance to crack propagation). If the interface is weak, the crack will cause the interface to fracture and result in failure of the composite. In this aspect, carbon nanotubes are better than traditional fibers (glass, carbon)due to their ability to inhibit nano- and micro-cracks. Hence, the knowledge and understanding of the nature and mechanics of load (stress) transfer between the nanotubes and polymer and properties of the interface is critical for manufacturing of mechanically enhanced CNT-polymer composites and will enable in tailoring of the interface for specific applications or superior mechanical properties. Broadly, the interfacial mechanics of CNT-polymer composites is appealing from three aspects: mechanics, chemistry, and physics. From a mechanics point of view, the important questions are:

1. The relationship between the mechanical properties of individual constituents, that is, nanotubes and polymer, and the properties of the interface and the composite overall.
2. The effect of the unique length scale and structure of the nanotubes on the property and behavior of the interface.
3. Ability of the mechanics modeling to estimate the properties of the composites for the design process for structural applications.

From a chemistry point of view, the interesting issues are:

1. The chemistry of the bonding between polymer and nanotubes, especially the nature of bonding (e.g., covalent or non-covalent and electrostatic).

2. The relationship between the composite processing and fabrication conditions and the resulting chemistry of the interface.

3. The effect of functionalization (treatment of the polymer with special molecular groups like hydroxyl or halogens) on the nature and strength of the bonding at the interface. From the physics point of view, researchers are interested in

   (a) The CNT-polymer interface serves as a model nano-mechanical or a lower dimensional system (1D) and physicists are interested in the nature of forces dominating at the nano-scale and the effect of surface forces (which are expected to be significant due to the large surface to volume ratio).

   (b) The length scale effects on the interface and the differences between the phenomena of mechanics at the macro (or meso) and the nano-scale.

2.7.2   SOME METHODS IN INTERFACE MODELING

Computational techniques have extensively been used to study the interfacial mechanics and nature of bonding in CNT-polymer composites. The computational studies can be broadly classified as atomistic simulations and continuum methods. The atomistic simulations are primarily based on molecular dynamic simulations (MD) and density functional theory (DFT) [66, 164, 165]. The main focus of these techniques was to understand and study the effect of bonding between the polymer and nanotubes (covalent, electrostatic or Van der Waals forces) and the effect of friction on the interface. The continuum methods extend the continuum theories of micromechanics modeling and fiber-reinforced composites (elaborated in the next section) to CNT-polymer composites [131, 166] and explain the behavior of the composite from a mechanics point of view.

On the experimental side, the main types of studies that can be found in literature are as follows:

1. Researchers have performed experiments on CNT-polymer bulk composites at the macroscale and observed the enhancements in mechanical properties (like elastic modulus and tensile strength) and tried to correlate the experimental results and phenomena with continuum theories like micromechanics of composites or Kelly Tyson shear lag model [73, 75, 167, 168].

2. Raman spectroscopy has been used to study the reinforcement provided by carbon nanotubes to the polymer, by straining the CNT-polymer composite and observing the shifts in Raman peaks [74, 169, 170].

3. In situ TEM straining has also been used to understand the mechanics, fracture and failure processes of the interface. In these techniques, the CNT-polymer composite (an electron transparent thin specimen) is strained inside a TEM and simultaneously imaged to get real-time and spatially resolved (1 nm) information [38].

2.7.3 NUMERICAL APPROACH

A MD model may serve as a useful guide, but its relevance for a covalent-bonded system of only a few atoms in diameter is far from obvious. Because of this, the phenomenological multiple column models that considers the interlayer radial displacements coupled through the Van der Waals forces is used. It should also be mentioned the special features of load transfer, in tension and in compression, in MWNT-epoxy composites studied by Schadler et al. [75] who detected that load transfer in tension was poor in comparison to load transfer in compression, implying that during load transfer to MWNTs, only the outer layers are stressed in tension due to the telescopic inner wall sliding (reaching at the shear stress 0.5 MPa [18, 40]), whereas all the layers respond in compression. It should be mentioned that NTCMs usually contain not individual, separated SWCNTs, but rather bundles of closest-packed SWCNTs [66] where the twisting of the CNTs produces the radial force component giving the rope structure more stable than wires in parallel. Without strong chemically bonding, load transfer between the CNTs and the polymer matrix mainly comes from weak electrostatic and Van der Waals interactions, as well as stress/deformation arising from mismatch in the coefficients of thermal expansion [171]. Numerous researchers [172] have attributed lower than predicted CNT-polymer composite properties to the availability of only a weak interfacial bonding. So, Frankland et al. [173] demonstrated by MD simulation that the shear strength of a polymer/nanotubes interface with only Van der Waals interactions could be increased by over an order of magnitude at the occurrence of covalent bonding for only 1% of the nanotubes carbon atoms to the polymer matrix. The recent force-field-based molecular-mechanics calculations

[69] demonstrated that the binding energies and frictional forces play only a minor role in determining the strength of the interface. The key factor in forming a strong bond at the interface is having a helical conformation of the polymer around the nanotubes; polymer wrapping around nanotubes improves the polymer-nanotubes interfacial strength, although configurationally thermodynamic considerations do not necessarily support these architectures for all polymer chains [5]. Thus, the strength of the interface may result from molecular-level entanglement of the two phases and forced long-range ordering of the polymer. To ensure the robustness of data reduction schemes that are based on continuum mechanics, a careful analysis of continuum approximations used in macromolecular models and possible limitations of these approaches at the nanoscale are additionally required that can be done by the fitting of the results obtained by the use of the proposed phenomenological interface model with the experimental data of measurement of the stress distribution in the vicinity of a nanotubes.

Meguid et al. [174] investigated the interfacial properties of carbon nanotubes (CNT) reinforced polymer composites by simulating a nanotubes pull-out experiment. An atomistic description of the problem was achieved by implementing constitutive relations that are derived solely from inter atomic potentials. Specifically, they adopt the Lennard-Jones (LJ) inter atomic potential to simulate a non-bonded interface, where only the van der Waals (vdW) interactions between the CNT and surrounding polymer matrix was assumed to exist. The effects of such parameters as the CNT embedded length, the number of van der Waals interactions, the thickness of the interface, the CNT diameter and the cut-off distance of the LJ potential on the interfacial shear strength (ISS) are investigated and discussed. The problem is formulated for both a generic thermo set polymer and a specific two-component epoxy based on diglycidyl ether of bisphenol A (DGEBA) and tri ethylene tetra mine (TETA) formulation. The study further illustrated that by accounting for different CNT capping scenarios and polymer morphologies around the embedded end of the CNT, the qualitative correlation between simulation and experimental pull-out profiles can be improved. Only van der Waals interactions were considered between the atoms in the CNT and the polymer implying a non-bonded system. The van der Waals interactions were simulated using the LJ potential, while the CNT was described using the Modified Morse potential. The results reveal that the ISS shows a linear dependence on the van der Waals interaction density and decays significantly with increasing nanotubes embedded length. The thickness of the interface was also varied and our results reveal that lower interfacial thicknesses favor higher ISS. When incorporating a 2.5 cut-off distance to the LJ potential, the predicted ISS shows an error of approximately 25.7% relative to a solution incorporating an infinite cut-off distance. Increasing the diameter of the CNT was found to increase the peak pull-out force approximately linearly. Finally, an examination of polymeric and CNT capping conditions showed that incorporating an end cap in the simulation yielded high initial pull-out peaks that better correlate with experimental findings. These findings have a direct bearing on the design and fabrication of carbon nanotubes reinforced epoxy composites.

Fiber pull-out tests have been well recognized as the standard method for evaluating the interfacial bonding properties of composite materials. The output of these tests is the force required to pullout the nanotubes from the surrounding polymer matrix and the corresponding interfacial shear stresses involved. The problem is formulated using a RVE which consists of the reinforcing CNT, the surrounding polymer matrix, and the CNT/polymer interface as depicted in (Fig. 2.7a, b) shows a schematic of the pull-out process, where x is the pullout distance and L is the embedded length of the nanotubes. The atomistic-based continuum (ABC) multi scale modeling technique is used to model the RVE. The approach adopted here extends the earlier work of Wernik and Meguid [175].

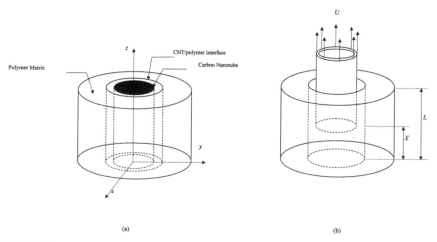

FIGURE 2.7   Schematic depictions of (a) the representative volume element and (b) the pull-out process.

The new features of the current work relate to the approach adopted in the modeling of the polymer matrix and the investigation of the CNT polymer interfacial properties as appose to the effective mechanical properties of the RVE. The idea behind the ABC technique is to incorporate atomistic inter atomic potentials into a continuum framework. In this way, the inter atomic potentials introduced in the model capture the underlying atomistic behavior of the different phases considered. Thus, the influence of the nano phase is taken into account via appropriate atomistic constitutive formulations. Consequently, these measures are fundamentally different from those in the classical continuum theory. For the sake of completeness, Wernik and Meguid provided a brief outline of the method detailed in their earlier work [174, 175].

The cumulative effect of the van der Waals interactions acting on each CNT atom is applied as a resultant force on the respective node, which is then resolved into its three Cartesian components. This process is depicted in Fig. 2.8 during each iteration of the pull-out process, the above expression is reevaluated for each van der Waals interaction and the cumulative resultant force and its three Cartesian components are updated to correspond to the latest pull-out configuration. Figure 2.9 shows a segment of the CNT with the cumulative resultant van der Waals force vectors as they are applied to the CNT atoms.

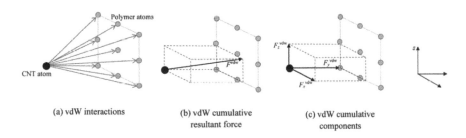

(a) vdW interactions   (b) vdW cumulative   (c) vdW cumulative
                                resultant force   components

FIGURE 2.8   The process of nodal van der Waals force application (a)van der Waals interactions on an individual CNT atom, (b) the cumulative resultant van der Waals force, and(c) the cumulative van der Waals Cartesian components.

Yang et al. [176] investigated the CNT size effect and weakened bonding effect between an embedded CNT and surrounding matrix were characterized using MD simulations. Assuming that the equivalent continuum

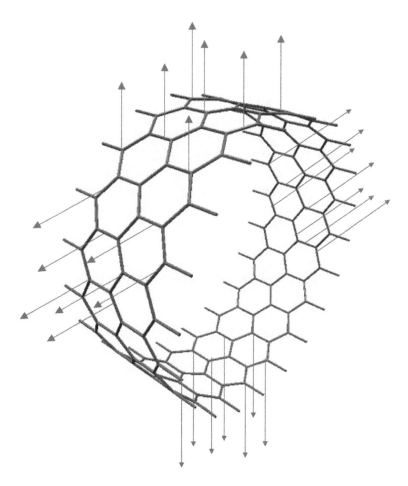

FIGURE 2.9   Segment of CNT with cumulative resultant van der Waals force vectors.

model of the CNT atomistic structure is a solid cylinder, the transversely isotropic elastic constants of the CNT decreased as the CNT radius increased. Regarding the elastic stiffness of the nanocomposite unit cell, the same CNT size dependency was observed in all independent components, and only the longitudinal Young's modulus showed a positive reinforcing effect whereas other elastic moduli demonstrated negative reinforcing effects as a result of poor load transfer at the interface. To describe the size effect and weakened bonding effect at the interface, a modified multiinclusion model was derived using the concepts of an effective CNT and effective matrix. During the scale bridging process incorporating the MD simulation results and modified multiinclusion model, we found that both the elastic modulus of the CNT and the adsorption layer near the CNT contributed to the size-dependent elastic modulus of the nano composites. Using the proposed multi scale-bridging model, the elastic modulus for nano composites at various volume fractions and CNT sizes could be estimated. Among three major factors (CNT waviness, the dispersion state, and adhesion between the CNT and matrix), the proposed model considered only the weakened bonding effect. However, the present multi scale framework can be easily applied in considering the aforementioned factors and describing the real nanocomposite microstructures. In addition, by considering chemically grafted molecules (covalent or non-covalent bonds) to enhance the interfacial load transfer mechanism in MD simulations, the proposed multi scale approach can offer a deeper understanding of the reinforcing mechanism, and a more practical analytical tool with which to analyze and design functional nano composites. The analytical estimation reproduced from the proposed multi scale model can also provides useful information in modeling finite element-based RVE s of nanocomposite microstructures for use in multifunctional design.

The effects of the inter phase and RVE configuration on the tensile, bending and Torsional properties of the suggested nanocomposite were investigated by Ayatollahi et al. [177]. It was found that the stiffness of the nanocomposite could be affected by a strong inter phase much more than by a weaker inter phase. In addition, the stiffness of the inter phase had the maximum effect on the stiffness of the nanocomposite in the bending loading conditions. Furthermore, it was revealed that the ratio of Le/Ln in RVE can dramatically affect the stiffness of the nanocomposite especially in the axial loading conditions.

For carbon nanotubes not well bonded to polymers, Jiang et al. [178] established a cohesive law for carbon nanotubes/polymer interfaces. The cohesive law and its properties (e.g., cohesive strength, cohesive energy) are obtained directly from the Lennard-Jones potential from the van der Waals interactions. Such a cohesive law is incorporated in the micromechanics model to study the mechanical behavior of carbon nanotubes-reinforced composite materials. Carbon nanotubes indeed improve the mechanical behavior of composite at the small strain. However, such improvement disappears at relatively large strain because the completely debonded nanotubes behave like voids in the matrix and may even weaken the composite. The increase of interface adhesion between carbon nanotubes and polymer matrix may significantly improve the composite behavior at the large strain [179].

Zalamea et al. [180] employed the shear transfer model as well as the shear lag model to explore the stress transfer from the outermost layer to the interior layers in MWCNTs. Basically, the interlayer properties between graphene layers were designated by scaling the parameter of shear transfer efficiency with respect to the perfect bonding. Zalamea et al. pointed out that as the number of layers in MWCNTs increases, the stress transfer efficiency decreases correspondingly. Shen et al. [181] examined load transfer between adjacent walls of DWCNTs using MD simulation, indicating that the tensile loading on the outermost wall of MWCNTs cannot be effectively transferred into the inner walls. However, when chemical bonding between the walls is established, the effectiveness can be dramatically enhanced. It is noted that in the above investigations, the loadings were applied directly on the outermost layers of MWCNTs; the stresses in the inner layers were then calculated either from the continuum mechanics approach [180] or MD simulation [181]. Shokrieh and Rafiee [159, 160] examined the mechanical properties of nano composites with capped single-walled carbon nanotubes (SWCNTs) embedded in a polymer matrix. The load transfer efficiency in terms of different CNTs' lengths was the main concern in their examination. By introducing an inter phase to represent the van der Waals interactions between SWCNTs and the surrounding matrix, Shokrieh and Rafiee [159, 160] converted the atomistic SWCNTs into an equivalent continuum fiber in finite element analysis. The idea of an equivalent solid fiber was also proposed by Gao and Li [182] to replace the atomistic structure of capped SWCNTs in the nanocomposites' cylindrical unit cell. The modulus of the equivalent solid was determined based on the atomistic structure of SWCNTs through molecular structure mechanics [183]. Subsequently, the continuum-based shear lag analysis was carried out to evaluate the axial stress distribution in CNTs. In addition, the influence of end caps in SWCNTs on the stress distribution of nanocomposites was also taken into account in their analysis. Tsai and Lu [184] characterized the effects of the layer number, intergraphic layers interaction, and aspect ratio of MWCNTs on the load transfer efficiency using the conventional shear lag model and finite element analysis. However, in their analysis, the interatomistic characteristics of the adjacent graphene layers associated with different degrees of interactions were simplified by a thin interphase with different moduli. The atomistic interaction between the grapheme layers was not taken into account in their modeling of MWCNTs. In light of the forgoing investigations, the equivalent solid of SWCNTs was developed by several researchers and then implemented as reinforcement in continuum-based nanocomposite models. Nevertheless, for MWCNTs, the subjects concerning the development of equivalent continuum solid are seldom explored in the literature. In fact, how to introduce the atomistic characteristics, that is, the interfacial properties of neighboring graphene layers in MWCNTs, into the equivalent continuum solid is a challenging task as the length scales used to describe the physical phenomenon are distinct. Thus, a multiscale based simulation is required to account for the atomistic attribute of MWCNTs into an equivalent continuum solid. In Lu and Tsai's study [184], the multiscale approach was used to investigate the load transfer efficiency from surrounding matrix to DWCNTs. The analysis consisted of two stages. First, a cylindrical DWCNTs equivalent continuum was proposed based on MD simulation where the pullout extension on the outer layer was performed in an attempt to characterize the atomistic behaviors between neighboring graphite layers. Subsequently, the cylindrical continuum (denoting the DWCNTs) was embedded in a unit cell of nano composites, and the axial stress distribution as well as the load transfer efficiency of the DWCNTs was evaluated from finite element analysis. Both single-walled carbon nanotubes (SWCNTs) and DWCNTs were considered in the simulation and the results were compared with each other.

An equivalent cylindrical solid to represent the atomistic attributes of DWCNTs was proposed in this study. The atomistic interaction of adjacent graphite layers in DWCNTs was characterized using MD simulation based on which a spring element was introduced in the continuum equivalent solid to demonstrate the interfacial properties of DWCNTs. Subsequently, the proposed continuum solid (denotes DWCNTs) was embedded in the matrix to form DWCNTs nano composites (continuum model), and the load transfer efficiency within the DWCNTs was determined from FEM analysis. For the demonstration purpose, the DWCNTs with four different lengths were considered in the investigation. Analysis results illustrate that the increment of CNTs' length can effectively improve the load transfer efficiency in the outermost layers, nevertheless, for the inner layers, the enhancement is miniature. On the other hand, when the covalent bonds between the adjacent graphene layers are crafted, the load carrying capacity in the inner layer increases as so does the load transfer efficiency of DWCNTs. As compared to SWCNTs, the DWCNTs still possess the less capacity of load transfer efficiency even though there are covalent bonds generated in the DWCNTs.

2.8   CONCLUDING REMARKS

Many traditional simulation techniques (e.g., MC, MD, BD, LB, Ginzburg-Landau theory, micromechanics and FEM) have been employed, and some novel simulation techniques (e.g., DPD, equivalent-continuum and self-similar approaches) have been developed to study polymer nano composites. These techniques indeed represent approaches at various time and length scales from molecular scale (e.g., atoms), to micro scale (e.g., coarse-grains, particles, monomers) and then to macroscale (e. g, domains), and have shown success to various degrees in addressing many aspects of polymer nano composites. The simulation techniques developed thus far have different strengths and weaknesses, depending on the need of research. For example, molecular simulations can be used to investigate molecular interactions and structure on the scale of 0.1–10 nm. The resulting information is very useful to understanding the interaction strength at nanoparticle-polymer interfaces and the molecular origin of mechanical improvement. However, molecular simulations are computationally very demanding, thus not so applicable to the prediction of mesoscopic structure and properties defined on the scale of 0.1–10 mm, for example, the dispersion of nanoparticles in polymer matrix and the morphology of polymer nano composites. To explore the morphology on these scales, mesoscopic simulations such as coarse-grained methods, DPD and dynamic mean field theory are more effective. On the other hand, the macroscopic properties of materials are usually studied by the use of mesoscale or macroscale techniques such as micromechanics and FEM. But these techniques may have limitations when applied to polymer nano composites because of the difficulty to deal with the interfacial nanoparticle-polymer interaction and the morphology, which are considered crucial to the mechanical improvement of nanoparticle-filled polymer nano composites. Therefore, despite the progress over the past years, there are a number of challenges in computer modeling and simulation. In general, these challenges represent the work in two directions. First, there is a need to develop new and improved simulation techniques at individual time and length scales. Secondly, it is important to integrate the developed methods at wider range of time and length scales, spanning from quantum mechanical domain (a few atoms) to molecular domain (many atoms), to mesoscopic domain (many monomers or chains), and finally to macroscopic domain (many domains or structures), to form a useful tool for exploring the structural, dynamic, and mechanical properties, as well as optimizing design and processing control of polymer nano composites. The need for the second development is obvious. For example, the morphology is usually determined from the mesoscale techniques whose implementation requires information about the interactions between various components (e.g., nanoparticle-nanoparticle and nanoparticle-polymer) that should be derived from molecular simulations. Developing such a multi scale method is very challenging but indeed represents the future of computer simulation and modeling, not only in polymer nano composites but also other fields. New concepts, theories and computational tools should be developed in the future to make truly seamless multi scale modeling a reality. Such development is crucial in order to achieve the longstanding goal of predicting particle-structure property relationships in material design and optimization.

The strength of the interface and the nature of interaction between the polymer and carbon nanotubes are the most important factors governing the ability of nanotubes to improve the performance of the composite. Extensive research has been performed on studying and understanding CNT-polymer composites from chemistry, mechanics and physics aspects. However, there exist various issues like processing of composites and experimental challenges, which need to be addressed to gain further insights into the interfacial processes.

KEYWORDS

- **Carbon Nanotubes**
- **Composite**
- **Micro Scale Methods**
- **Modeling**
- **Molecular Dynamics**
- **Numerical Approach**

CHAPTER 3

INTER-ATOMIC RELATIONS IN CARBON NANOTUBES

CONTENTS

3.1   INTRODUCTION

At the beginning of this considerations, let introduce basic relations describing carbon–carbon (C–C) interactions. They characterize physical behavior of carbon nanotubes, being in fact atomistic structures, and are fundamental in the further transformation from molecular dynamics relations to continuum (shell) mechanics. The structure of nanotubes is obtained by conformational mapping of a graphene sheet onto a cylindrical surface. The nanotubes radius is estimated by using the relation

$$R = r_0 \frac{\sqrt{3(m^2 + n^2 + mn)}}{2\pi} \tag{1}$$

where $r_0 = 0.141$ nm is the carbon-carbon distance. The integer's n and m denote the number of unit vectors $\mathbf{a}_1$ and $\mathbf{a}_2$ along two directions in the honeycomb crystal lattice of graphene. If $m = 0$, they are called "zigzag" nanotubes; if $n = m$, they are called "armchair" nanotubes. For any other values of n and m, the nanotubes are called "chiral" because the chains of atoms spiral around the tube axis instead of closing around the circumference. To capture the essential feature of chemical bonding in graphite, Brenner (1990) established an inter atomic potential (called as the REBO potential) for carbon in the following form

$$V\left(r_{ij}\right) = V_R\left(r_{ij}\right) - B_{ij} V_A\left(r_{ij}\right) \tag{2}$$

where r_{ij} is the distance between the atoms i and j, V_R and V_A are the repulsive and attractive pair terms (i.e., depending only on r_{ij}), and are given by

$$V_R(r) = \frac{D_{(e)}}{S-1} exp\left[-\sqrt{2S}\beta(r - R)\right] V_A(r) = \frac{D_{(e)}S}{S-1} exp\left[-\sqrt{\frac{2}{S}}\beta(rR)\right] \tag{3}$$

In the above expression, the cut-off function is assumed to be equal to 1 to avoid a dramatic increase in the interatomic force.

The parameter B_{ij} in Eq. (2) represents the multi body coupling between the bond from the atoms i and j and the local environment of the atom i, and is given by

$$B_{ij} = \left[1 + \sum_{k(\neq j, j)} G(\theta_{ijk})\right]^{-\delta} \tag{4}$$

where θ_{ijk} is the angle between bonds $i-j$ and $i-k$, and the function G is given by

$$G(\theta) = a_0 \left[1 + \frac{c_0^2}{d_0^2} + \frac{c_0^2}{d_0^2 + (1+\cos\theta)^2} \right] \tag{5}$$

and the term B_{ij} is expressed in the symmetric form

$$\overline{B_{iJ}} = \frac{1}{2}(B_{ij} + B_{ij}) \tag{6}$$

The set of material parameters is adopted here as follows

$$R = 0.139nm \quad \delta = 0.5a_0 = 0.00020813$$
$$c_0 = 330d_0 = 3.5$$

In contrast to the REBO potential function, in which the bond stretch and bond angle are coupled in the potential, Belytschko et al. [185] proposed the modified Morse potential function, which can be expressed as the sum of energies that are associated with the variance of the bond length V stretch, and the bond angle Vangle, that is,

$$V = V_{stretch} + V_{angle} \tag{7}$$

$$V_{stretch} = D_e \left\{ \left[1 - \exp(-\beta(r - r_0)) \right]^2 - 1 \right\} \tag{8}$$

$$V_{angle} = \frac{1}{2}k_\theta(\theta - \theta_0)^2 \left[1 + k_{sertic}(\theta - \theta_0)^4 \right] \tag{9}$$

The material constants are following

$$r_0 = 1.421.10^{-10} m \quad D_e = 9.10^{-19} Nm$$
$$\beta = 1.8.10^{10} m^{-1} \theta = 120°$$
$$k_\theta = 0.9.10^{-18} \frac{Nm}{rad^2} k_{sertic} = 0.754rad^{-4}$$

By differentiating Eq. (2) or (8), the stretching force of atomic bonds is obtained. The force variations with the bond length are almost the same while the bond angle is kept constant as $2\pi/3$. However, for the REBO potential, the force varies with the bond angle variations, whereas for the modified Morse potential it is always constant. Thus, the inflection point (force peak) is not constant for the REBO potential. As it is reported, both bond lengths and bond angles vary as CNTs are stretched. Therefore, in our numerical model, it is necessary to consider two possible formulations of inter atomic potentials to analyze and compare the influence of those effects on the non-linear behavior and fracture strain.

From the inter atomic potentials that are shown in Eqs. (2)and(8), the stretching force that results from the bond elongation and the twisting moment that results from the bond angle variation can be calculated as follows

$$F(r_i) = \frac{\partial V}{\partial r_i} \quad M(\theta) = \frac{\partial V}{\partial \theta} \tag{10}$$

Figure 3.1 compares the inter atomic stretching force for the REBO and modified Morse potentials in the tensile regime, whereas (Fig. 3.1) shows the bond angle moment for the REBO and modified Morse potentials. Let us note that they presentnon-linear behavior.

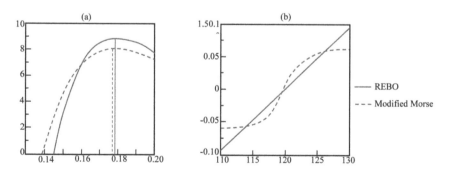

FIGURE 3.1   Tensile (a) force and (b) moment distributions.

3.2   CONTINUUM SHELL MODEL FOR SWCNT

To describe mechanical behavior of carbon nanotubes in the form of classical shell relations, it is necessary to introduce the equivalent Young modulus E and the equivalent shell thickness h(Fig. 3.2).

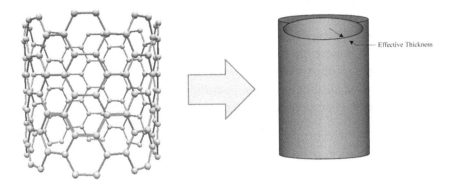

FIG. 3.2   Continuum model of carbon nanotubes structures.

In general, three different approaches are possible:
- To assume the above-mentioned values; for instance Yakobson et al. [186] suggested that the effective Young modulus is 5.5 TPa and the wall thickness of the carbon nanotubes is 0.066 nm based on their SWCNT buckling results obtained by MD simulation and the continuum mechanics shell model.
- To fit the results to atomistic simulation results of tension rigidity $E_A = E_h/(1-v_2)$ and bending rigidity $EI = Eh3/[12(1-v_2)]$, where v is the Poisson ratio; this gives [187]:

$$h = \sqrt{12\frac{EA}{EI}} = 3\sqrt{2(\frac{\partial V}{\partial \cos \theta_{ijk}})_0 (\frac{\partial^2 V}{\partial r_{ij}^2})_0^{-1}} \tag{11}$$

- To determine analytically tensile, shear, bending and Torsional rigidities directly from inter atomic potential, and therefore avoid any fitting not-well defined elastic modulus and thickness [188]. This atomistic-based shell theory gives the relation among the increments of second Piola-Kirchhoff stress **T**, moment **M**, Green strain **E**, and curvature **K** as

$$T = L:E+H:K \quad M = H:E+S:K \tag{12}$$

where **L**, **H** and **S** are the fourth-order rigidity tensors obtained analytically from the inter atomic potential.

Using the above-mentioned approaches, it is possible to evaluate buckling loads of cylindrical shells, analyze their post-buckling behavior, etc., just employing the methods of analysis well-known for shell problems. In this way, it is possible to describe the behavior of equivalent carbon single-walled or multi-walled carbon nanotubes. However, there is always an open question dealing with the accuracy and correctness of such approaches. Some of those problems will be discussed in the next section.

3.3 PROBLEMS ENCOUNTERED IN CONTINUOUS CYLINDRICAL MODELING

Peng et al. [189] determined the order of error for approximating single-walled carbon nanotubes by a thin shell. The ratio of atomic spacing r_0 to the single-walled carbon nanotubes radius R *(i.e., r_0/R)*, is used to identify the order of error. They considered the structural response of single-walled carbon nanotubes subject to tension (or compression), torsion, bending and internal (or external) pressure. They proved that only for the order of error equal to 40%–(5.5) armchair single walled carbon nanotubes can be modeled as thin shells with constant thickness and isotropic mechanical properties. The extensions of the above results were presented by Wu et al. [188]. The authors defined degrees of: anisotropy, non-linearity and coupling, that is, down to the radius of single-walled carbon nanotubes at which the tension/bending coupling becomes negligible in constitutive relation (3.2). Numerical modeling of single-walled carbon nanotubes constitutes a separate class of problems. Some of them were discussed by Kalmakarov et al. [190]. This cited work presents also the comparison of Young's modulus and the equivalent thickness predicted with the use of various theories.

3.4 ANALYTICAL TECHNIQUE BASED ON ASYMPTOTIC HOMOGENIZATION

The proposed analytical technique assumes that single-walled carbon nanotubes (SWCNT) can be modeled as a cylindrical network shell with a hexagonal periodicity cell. The shell in turn can be assumed as an inhomogeneous thin three-dimensional layer with zero elastic properties in areas of perforation [191]. Due to their periodic configuration, SWCNTs can be modeled using asymptotic homogenization techniques [192–194].

According to this method, two levels of spatial variables are considered, one for the description of the media at the micro scale, and the other variable for the global changes in the physical field of interest at the macroscale. The partial differential equations of the problem have coefficients represented by periodic functions in the form of $A(x/\delta) = A(\xi)$. The corresponding boundary value problem is treated by asymptotically expanding the solution in terms of the characteristic small parameter, d, thus making them dependent on both the slow (macroscopic) variable x and the rapidly oscillating (microscopic) variable $\xi = x/\delta$. For the analytical treatment, the carbon–carbon (C–C) bond is represented by l, and the bond between adjacent carbon atoms in a hexagonal ring is represented by a circular bar of diameter δ.

3.4.1 ASYMPTOTIC HOMOGENIZATION OF A GENERAL COMPOSITE SHELL

The formulation used in the analytical method is similar to that used in the determination of the effective coefficients for a composite thin layer (shell) with rapidly varying thickness [190, 192–195]. In this section, only a brief overview will be given. Consider a general regularly inhomogeneous three-dimensional thin layer obtained by repeating a small unit cell Ω_δ as shown in Fig. 3.3. Here, $(\alpha_1, \alpha_2$ and $\gamma)$ represent an orthogonal coordinate system, such that the coordinate lines a_1 and a_2 coincide with the main curvature lines of the mid-surface of the layer, and c is the coordinate axis normal to the mid-surface.

It is assumed that the thickness of the layer and the tangential dimensions of the periodicity cell of the structure are small in comparison with the dimensions of the composite layer in whole. These small dimensions are characterized by a small dimensionless parameter δ, which defines both the scale of in homogeneity in the medium and the thickness of the layer. The unit cell Ω_δ is defined by the following inequalities, (see Fig. 3.3):

$$-\frac{\delta h_1}{2} < \alpha_1 < \frac{\delta h_1}{2}, -\frac{\delta h_2}{2} < \alpha_2 < \frac{\delta h_2}{2}, \gamma^- < \gamma < \gamma^+ \qquad (13)$$

where

$$\gamma^\pm = \pm \frac{\delta}{2} \pm \delta F^\pm \left(\frac{\alpha_1}{\delta h_1}, \frac{\alpha_2}{\delta h_2} \right).$$

(14)

Here, functions $F^\pm$ define the profiles of the upper (S^+) and lower (S^-) surfaces of the layer. $F^\pm$ are assumed to be piecewise-smooth periodic functions in variables α_1 and α_2 with a periodicity defined by unit cell Ω_δ. The solution of this 3D elasticity problem is represented in terms of asymptotic expansions in powers of the small parameter d. The constitutive relations of the equivalent (homogenized) shell are obtained in terms of the stress resultants ($N_{\alpha\beta}$), moment resultants ($M_{\alpha\beta}$) and the mid-surface strains ($\varepsilon_{\alpha\beta}$) and curvatures($k_{\alpha\beta}$), see [193]. They are given as:

$$N_{\alpha\beta} = \delta \left\langle b_{\alpha\beta}^{\lambda\mu} \right\rangle \varepsilon_{\lambda\mu} + \delta^2 \left\langle b_{\alpha\beta}^{*\lambda\mu} \right\rangle k_{\lambda\mu}$$

(15)

$$M_{\alpha\beta} = \delta^2 z b_{\alpha\beta}^{\lambda\mu} \varepsilon_{\lambda\mu} + \delta^3 z b_{\alpha\beta}^{*\lambda\mu} k_{\lambda\mu}$$

(16)

Throughout this work, it is assumed that Greek indices α, β, λ, etc. take values 1 and 2, whereas Latin indices, i, j, k, etc. vary from 1 to 3. The quantities $<b^{\lambda\mu}_{\alpha\beta}>$, $<b^{*\lambda\mu}_{\alpha\beta}>$, $<zb^{\lambda\mu}_{\alpha\beta}>$, and$<zb^{*\lambda\mu}_{\alpha\beta}>$ are called the effective elastic parameters.

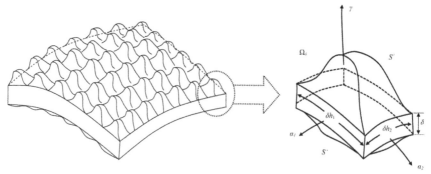

FIGURE 3.3   Unit cells of the inhomogeneous cylindrical layer.

Coefficients of the homogenized shell and are obtained through integration over the entire unit cell Ω_δ, according to

$$\left\langle f(\xi_1, \xi_2, \xi_3) \right\rangle = \frac{1}{|\Omega|} \int_\Omega \int^0 f(\xi_1, \xi_2, z) d\xi_1 d\xi_2 dz$$

(17a)

Here, ξ_1, ξ_2 and z are the macroscopic variables and are defined as:

$$\xi_1 = \frac{\alpha_1 A_1}{\delta h_1}$$

$$\xi_2 = \frac{\alpha_2 A_2}{\delta h_2}$$

$$z = \frac{\gamma}{\delta} \tag{17b}$$

where $A_1(a_1, a_2)$ and $A_2(a_1, a_2)$ are the coefficients of the first quadratic form of the mid-surface layer, and $|\Omega|$ is the volume of the unit cell in coordinates ξ_1, ξ_2 and z.

Before Eqs (15) and (16) are applied, the local functions $<b^{\lambda\mu}{}_{\alpha\beta}>$, $<b*^{\lambda\mu}{}_{\alpha\beta}>$ must be determined from the following unit cell problems [193]:

$$\frac{1}{h_\beta}\frac{\partial b^{\lambda\mu}_{i\beta}}{\partial\varsigma_\beta} + \frac{\partial b^{\lambda\mu}_{i3}}{\partial Z} = 0 \tag{18}$$

$$\frac{1}{h_\beta} n^{\pm}_\beta b^{\lambda\mu}_{i\beta} + n^{\pm}_3 b^{\lambda\mu}_{i3} = 0 \; atz = z^{\pm} \tag{19}$$

$$\frac{1}{h_\beta}\frac{\partial b^{*\lambda\mu}_{i\beta}}{\partial\varsigma_\beta} + \frac{\partial b^{*\lambda\mu}_{i3}}{\partial z} = 0 \tag{20}$$

$$\frac{1}{h_\beta} n^{\pm}_\beta b^{*\lambda\mu}_{i\beta} + n^{\pm}_3 b^{*\lambda\mu}_{i3} = 0 \; atz = z^{\pm} \tag{21}$$

Here, $n^+_i(n^-_i)$ are the components of the outward (inward) unit normal vector corresponding to the surface, $z = z^+(z = z^-)$ and defined in the coordinate system ξ_1, ξ_2, z. Problems (18) and (19) with appropriate boundary conditions (20) and (21) and periodicity conditions in tangential directions are solved entirely on the domain of the unit cell and are called "unit cell problems."

In fact, the local functions $<b^{\lambda\mu}{}_{\alpha\beta}>$, $<b*^{\lambda\mu}{}_{\alpha\beta}>$ are not solved directly from Eqs. (18)and(20). Instead, the following definitions:

$$b^{*lm}_{ij} = \frac{1}{h_\beta} C_{ijn\beta}\frac{\partial U^{lm}_n}{\partial\varsigma_\beta} + C_{ijn3}\frac{\partial U^{lm}_n}{\partial z} + C_{ijlm} \tag{22}$$

$$b_{ij}^{*lm} = \frac{1}{h_\beta} c_{ijn\beta} \frac{\partial V_n^{lm}}{\partial \varsigma_\beta} + c_{ijn3} \frac{\partial V_n^{lm}}{\partial z} + zc_{ijlm} \qquad (23)$$

are used to relate the local functions with the yet unknown functions $U_n^{lm}(\xi_1, \xi_2, z)$, $V_n^{lm}(\xi_1, \xi_2, z)$ and the elastic coefficients of the material, c_{ijlm}. These functions are periodic in ξ_i (with respective period A_i) but not in z.

Hence, Eqs.(22)and(23) are first substituted in Eqs. (18)–(21) and then the functions $U^{lm}_{\ n}$ and $V^{lm}_{\ n}$ are determined. These are in turn back substituted in Eqs. (22)and(23) to obtain the local functions $<b^{\lambda\mu}_{\ \alpha\beta}>$ and $<b^{*\lambda\mu}_{\ \alpha\beta}>$ and finally these are used to determine the effective elastic coefficients by averaging over the volume of the unit cell according to Eq. (17a). It should be noted, as can also be observed from Eqs. (15)and(16) that there is a correspondence between the effective elastic coefficients and the extensional, A_{ij}, coupling, B_{ij} and bending D_{ij}, coefficients familiar from the classical composite laminate theory [196]. These are

$$\begin{bmatrix} A & B \\ B & D \end{bmatrix} = \begin{bmatrix} \delta\langle b_{11}^{11}\rangle & \delta\langle b_{11}^{22}\rangle & \delta\langle b_{11}^{12}\rangle & \delta^2\langle zb_{11}^{11}\rangle & \delta^2\langle zb_{11}^{22}\rangle & \delta^2\langle zb_{11}^{12}\rangle \\ \delta\langle b_{11}^{22}\rangle & \delta\langle b_{22}^{22}\rangle & \delta\langle b_{22}^{12}\rangle & \delta^2\langle zb_{11}^{22}\rangle & \delta^2\langle zb_{22}^{22}\rangle & \delta^2\langle zb_{22}^{12}\rangle \\ \delta\langle b_{11}^{12}\rangle & \delta\langle b_{22}^{12}\rangle & \delta\langle b_{12}^{12}\rangle & \delta^2\langle zb_{11}^{12}\rangle & \delta^2\langle zb_{22}^{12}\rangle & \delta^2\langle zb_{12}^{12}\rangle \\ \delta^2\langle b_{11}^{*11}\rangle & \delta^2\langle b_{11}^{*22}\rangle & \delta^2\langle b_{11}^{*12}\rangle & \delta^3\langle zb_{11}^{*11}\rangle & \delta^3\langle zb_{11}^{*22}\rangle & \delta^3\langle zb_{11}^{*12}\rangle \\ \delta^2\langle b_{11}^{*22}\rangle & \delta^2\langle b_{22}^{*22}\rangle & \delta^2\langle b_{22}^{*12}\rangle & \delta^3\langle zb_{11}^{*22}\rangle & \delta^3\langle zb_{22}^{*22}\rangle & \delta^3\langle zb_{22}^{*12}\rangle \\ \delta^2\langle b_{11}^{*12}\rangle & \delta^2\langle b_{22}^{*12}\rangle & \delta^2\langle b_{12}^{*12}\rangle & \delta^3\langle zb_{11}^{*12}\rangle & \delta^3\langle zb_{22}^{*12}\rangle & \delta^3\langle zb_{12}^{*12}\rangle \end{bmatrix} \qquad (24)$$

The method of asymptotic homogenization has been successfully applied to analyze a number of practically important configurations of composite and smart composite shells and plates [190, 192–195]. This technique is used here to derive the engineering constants and constitutive relations pertaining to single-walled carbon nanotubes.

3.4.2 ASYMPTOTIC TREATMENT OF CYLINDRICAL NETWORK SHELL WITH PERIODIC STRUCTURE

As mentioned earlier, a SWCNT is modeled as a three-dimensional inhomogeneous cylindrical shell. The nanotubes shell is strictly periodic and void of any topological defects such as Stone-Wales defects and vacancy defects; the radius of curvature of the shell is much larger than the shell thickness; terms of order higher than O (δ^2) are neglected in the pertinent asymptotic assumptions [193, 195]; the constituent nanotubes material is homogeneous; the bars representing the chemical bonds are cylindrical. The nanotubes shell has a uniform thickness d,which implies that $F^+ \equiv F^- \equiv 0$ (see Fig. 3.4). Further, it is assumed that the region of space of perforations (which does not contain any carbon-carbon bonds) is assigned zero material properties [197].

The effective coefficients of this structure are determined on the basis of the solution of the local problems (the Eqs. (18)–(23)) on the individual unit cell. Suppose that the unit cell of the network shell is formed by N bars, such that the j-th bar ($j = 1, 2, 3..., N$), which is made of isotropic material with Young's modulus E_j and Poisson's ratio v_j, subtends an angle φ_j with coordinate line a_1. Local problems (the Eqs.(18)–(21)) can be solved for each bar of the unit cell separately (see Fig. 3.4), and the effective stiffness of the entire structure can be determined by superposition using Eq. (17a). We note that by using this method, we accept the error incurred at the regions of overlap of the bars. However, this error is highly localized and does not contribute significantly to the integral over the unit cell. Following this procedure for the case of a SWCNT the effective coefficients are easily determined from Eqs (18), (19), (22) and (23) the averaging Eq. (17), and the angle φ_j between the j-th element of the unit cell and the α_1 axis. The results are:

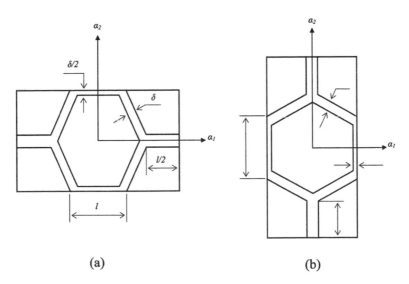

(a)   (b)

FIGURE 3.4 Periodicity cells of SWCNT (a) armchair and (b) zig-zag.

$$\left\langle b_{\alpha\beta}^{\lambda\mu} \right\rangle = \sum_{j=1}^{N} E_j B_j^{\alpha\beta\lambda\mu} \gamma_j \tag{25}$$

$$\left\langle z b_{\alpha\beta}^{\lambda\mu} \right\rangle = \left\langle b_{\alpha\beta}^{*\lambda\mu} \right\rangle = 0 \tag{26}$$

$$\left\langle z b_{\alpha\beta}^{\lambda\mu} \right\rangle = \sum_{j=1}^{N} E_j (B_j^{\alpha\beta\lambda\mu} \gamma_j + \frac{C_j^{\alpha\beta\lambda\mu}}{1+v_j}) \frac{\gamma_j}{16} \tag{27}$$

Here, the functions $B^{\alpha\beta\lambda\mu}{}_j$ and $C^{\alpha\beta\lambda\mu}{}_j$ depends on the index combination α, β, κ, λ are given as:

$$B_j^{1111} = cos^4\varphi_j C_j^{1111} = cos^2\varphi_j sin^2\varphi_j \tag{28}$$

$$B_j^{2222} = sin^4\varphi_j C_j^{2222} = sin^2\varphi_j cos^2\varphi_j \tag{29}$$

$$B_j^{1212} = cos^2\varphi_j sin^2\varphi_j C_j^{1212} = \frac{1}{4}(cos^4\varphi_j + sin^4\varphi_j - 2cos^2\varphi_j sin^2\varphi_j) \tag{30}$$

$$B_j^{1112} = B_j^{2211} = cos^2\varphi_j sin^2\varphi_j C_j^{1112} = C_j^{2211} = -cos^2\varphi_j sin^2\varphi_j \tag{31}$$

$$B_j^{1112} = B_j^{1211} = cos^3\varphi_j sin\varphi_j C_j^{1112} = C_j^{1211} = \frac{1}{2} cos\varphi_j sin^3\varphi_j - cos^3\varphi_j sin\varphi_j) \tag{32}$$

$$B_j^{1222} = B_j^{2212} = cos\varphi_j \ sin^3 \ \varphi_j C_j^{1222} = C_j^{2212} = \frac{1}{2}(cos^3 \ \varphi_j \ sin\varphi_j - cos\varphi_j \ sin^3 \ \varphi_j) \tag{33}$$

It is noteworthy to mention here that the effective $<b^{\lambda\mu}{}_{\alpha\beta}>$ coefficients are independent of the shape of the cross section of the bars and only depend on their volume fraction (i.e., their cross-sectional area). On the contrary, the effective $<b*^{\lambda\mu}{}_{\alpha\beta}>$ coefficients are necessarily dependent on the shape of the cross-section due to their dependence on the transverse coordinate.

Referring to Fig. 3.4a, the stress and moment resultants are determined as follows [197]:

$$N_{11} = \delta^2 \frac{E}{1} \frac{\pi}{16\sqrt{3}} (3\varepsilon_{11} + \varepsilon_{22}) \tag{34}$$

$$N_{22} = \delta^2 \frac{E}{1} \frac{\pi}{16\sqrt{3}} (\varepsilon_{11} + 3\varepsilon_{22}) \tag{35}$$

$$N_{12} = \delta^2 \frac{E}{1} \frac{\pi}{16\sqrt{3}} \varepsilon_{12} \tag{36}$$

$$M_{11} = \delta^3 \frac{E}{(1+v)l} \frac{\pi\sqrt{3}}{768} \left[(4+3v)k_{11} + vk_{22} \right] \tag{37}$$

$$M_{22} = \delta^3 \frac{E}{(1+v)l} \frac{\pi\sqrt{3}}{768} \left[vk_{11} + (4+3v)k_{22} \right] \tag{38}$$

$$M_{12} = \delta^3 \frac{E}{(1+v)l} \frac{\pi\sqrt{3}}{768} \left[(v+1)k_{12} \right] \tag{39}$$

The results pertinent to the unit cell of Fig. 3.4 can be obtained by interchanging indices 1 for 2, in Eqs (29) and (30). The calculation of the effective elastic and shear moduli follow from Eqs. (29)and(30) and are given below.

3.4.3 NUMERICAL TECHNIQUE BASED ON FINITE ELEMENT IMPLEMENTATION

For the numerical study, the carbon nano tube is replaced by an equivalent structural model. The nanotubes can be visualized as carbon atom network shell, obtained by rolling up a graphene sheet. As the carbon atoms in both the graphene sheet and the rolled-up nanotubes are held together by covalent bonds of characteristic bond length, the carbon atoms can be viewed as material points which are connected by load carrying beam elements [183]. Co-relations between molecular mechanics and structural mechanics are used to establish the equivalency between the two models. The structural model is then analyzed using the developed finite element computational procedure. The elastic response of structural model is then used to calculate the properties of continuum tube (shell).

3.4.4 CORRELATION BETWEEN STRUCTURAL AND MOLECULAR MECHANICS

At the molecular level, the interaction between individual carbon atoms can be described using the force fields of the corresponding nucleus-nucleus and electron-nucleus interactions [198]. If electrostatic interactions are neglected, the total steric potential energy (U total), which characterizes the force field, can be obtained as the sum of energies due to valence (or bonded) and non-bonded interactions [199], given as

$$U_{total} = \sum U_r + \sum U_\theta + \sum U_\varnothing + \sum U_{vdw} \tag{40}$$

Here, U_r, U_θ, U_φ, U_{vdw} correspond to energy associated with bond stretch interactions, bond angle bending, torsion (dihedral and out of plane) and van der Waals forces (non-covalent). Figure 3.5a, illustrates the various interatomic interactions at the molecular level. Several harmonic and non-harmonic potential functions have been proposed to describe the interatomic interactions of carbon atoms [199–202]. Gelin [203] suggested that for small deformations simple harmonic approximations are sufficient to describe the potential energy of the force fields. Assuming that

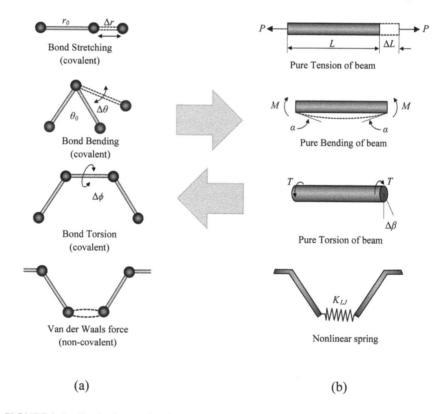

(a)   (b)

FIGURE 3.5   Equivalence of molecular mechanics and structural mechanics for covalent and non-covalent interactions (a) molecular mechanics model and (b) structural mechanics model.

the covalent interactions between carbon atoms can be represented using simple harmonic functions, the Vibrational potential energies due to interactions between covalently bonded carbon atoms can be represented by Eqs. (32)–(34):

$$\text{Bond stretching energy,}\ U_r = \frac{1}{2}k_r(r-r_0)^2 = \frac{1}{2}k_r(\Delta r)^2 \tag{41}$$

$$\text{Bond bending energy,}\ U_\theta = \frac{1}{2}k_\theta(\theta-\theta_0)^2 = \frac{1}{2}k_\theta(\Delta\theta)^2 \tag{42}$$

$$\text{Total torsional energy,}\ U_\varnothing = \frac{1}{2}k_\varnothing(\Delta\varnothing)^2 \tag{43}$$

Here, r_0 and h_0 refer to the undeformed interatomic distance and undeformed bond angle for a bond (see Fig. 31a), and the quantities r and h refer to the distance and bond-angle after deformation. Consequently, the terms Δr, $\Delta\theta$, $\Delta\phi$ correspond to change in bond-length, bond angle and dihedral angle, respectively. The terms k_r, k_θ and k_φ represent the force constants associated with stretching, bending and torsion, respectively, of the chemical bond. The carbon atoms in the nanotubes are held together by covalent bonds of characteristic bond length and bond angles, and the corresponding molecular forces constrain any displacement of individual atoms [183]. The non-covalent interactions like van der Waals forces can be adequately described using Lennard-Jones potential [204]. The corresponding energy is given by

$$V_{LJ} = 4\varepsilon\left[(\frac{\sigma}{r})^{12} - (\frac{\sigma}{r})^6\right] \tag{44}$$

In Eq. (3.44) the terms r (in nm) and ε (in kJ/mol) are defined as the Lennard-Jones parameters. They are material specific and determine the nature and strength of the interaction. The term r corresponds to the distance between the interacting particles. A typical curve of the Lennard-Jones potential is given in Fig. 3.32.

Due to the nature of the molecular force fields between two atoms, they can be treated as forces acting between two junctions (or material points) that are separated by structural beam or spring elements. Thus, the lattice of the carbon nanotubes can be considered as a three dimensional hexagonal network of beam (covalent) and spring (non-covalent) elements.

Figure 3.6 illustrates the correlation between beam elements and the molecular forces between bonded carbon atoms. To determine the force constants pertaining to the covalent interactions one could equate the potential energies of individual bonds with their corresponding beam model [183].

The beam elements representing the bond are assumed to be isotropic with length L, cross sectional area A, and moment of inertia I. The strain energy under pure axial load P, (pure tension) is given by:

$$U_p = \int_0^L \frac{P^2}{2EA} \, dL = \frac{EA}{2L}(\Delta L)^2 \tag{45}$$

The strain energy of beam element subjected to pure bending moment M, is given by:

$$U_M = \int_0^L \frac{M^2}{2EL} \, dL = \frac{EI}{2L}(2\alpha)^2 \tag{46}$$

Similarly, the strain energy of the beam element under a pure twisting moment T, is given by

$$U_T = \int_0^L \frac{T^2}{2GL} \, dL = \frac{GI}{2L}(\Delta\beta)^2 \tag{47}$$

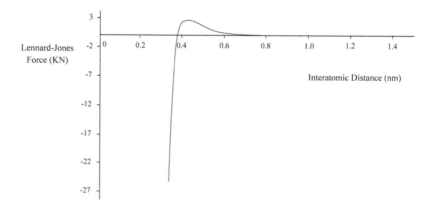

FIGURE 3.6 Variation of the Lennard-Jones force with interatomic distance of carbon atoms.

In the above equations, the terms ΔL, α and $\Delta \beta$ are the axial deformation, bend angle and twist angle, respectively. Eqs (41)–(43) and Eqs (45)–(47) represent the same quantities in two different systems (molecular and structural) so they can be equated, thus establishing a link between the two systems. Also, assuming the terms ΔL, $\Delta \theta$ and $\Delta \varphi$ are equivalent to their structural mechanics counterparts, that is, Δr, 2α and $\Delta \beta$, respectively, one obtains the tensile stiffness (EA), cross-sectional bending stiffness (EI) and torsional rigidity (GJ) of the structural model in terms of the molecular mechanics force constants k_r, k_θ and k_φ [183]. These are:

$$\frac{EA}{L} = k_r, \frac{EI}{L} = k_\theta, \frac{GJ}{L} = k_\varnothing \tag{48}$$

In the structural system, the van der Waals forces due tonon-covalent interactions are assumed to be mimicked by spring elements. The force acting in such a spring element can be obtained by differentiating Eq. (44) and is given by

$$F_{IJ} = \frac{dV(r)}{dr} = 4\frac{\varepsilon}{r}\left[-12(\frac{\sigma}{r})^{12} + 6(\frac{\sigma}{r})^{6}\right] \tag{49}$$

In summary, the parameters in Eqs (48) and (49) are used to model the molecular behavior in the structural model. In case of single-walled carbon nanotubes (with only covalent bonds) the parameters in Eq. (48) are sufficient to describe the structural model (with beam elements). In case of multi-walled carbon nanotubes, the van der Waals interactions between carbon atoms in different concentric tubes must also be considered, and hence Eq. (49) is used as well.

3.4.5 MODELING OF NANOTUBES USING FINITE ELEMENT MODEL

The structural model discussed above describes a nanotubes model using beam and spring elements. Furthermore, the carbon atoms will be denoted by nodes at appropriate locations. The global coordinates of carbon atoms in a nano tube could be traced by applying a mere transformation on the corresponding locations of the carbon atoms on the graphite plane [205]. Subsequently, the finite element models pertaining to various types (single-, double- and multi-walled) and configurations (zig-zag and arm-chair) of the carbon nanotubes have been developed. The covalent interactions are modeled using three-dimensional beam elements. These beam elements have six degrees of freedom at each node: translations in nodal x, y, z and rotations about the nodal x, y, z. they are capable of uniaxial tension or compression along with Torsional and bending deformations. Table 3.1 summarizes the geometric and material properties of the elastic beam element,which are input to FE model. The spring elements corresponding to non-covalent interactions between carbon atoms are modeled using a non-linear spring element, which is capable of uniaxial tension or compression. The element has three degrees of freedom at each node (translations in nodal x, y and $z$axes) and is specified by two nodes and a non-linear force-displacement relationship. The elastic properties of the beam elements (assumed to have a diameter db and length l) are assumed to be isotropic and obtained from Eq. (48). They are:

$$\text{The bond diameter, } d_b = 4\sqrt{\frac{k_\theta}{k_r}} \tag{50}$$

$$\text{Young's modulus of the elastic beam, } E = \frac{k_r^2 L}{4\pi k_\theta} \tag{51}$$

TABLE 3.1.   Input Sectional and Properties of the Beam Element

Cross-sectional area, A	1.68794A²
Moment of inertia $I_{yy}=I_{zz}=I$	0.22682A⁴
Polar moment of inertia I_{xx}	0.453456A⁴
Elastic modulus of beam element, E	5.488×10^{-8} N/A2
Shear modulus of beam element, G	8.711×10^{-9} N/A2

The present investigation assumes typical values for the force field constants, k_r, k_θ and k_φ as 6.52×10^{-7} N/nm, 8.76×10^{-10}N nm/rad² and 2.78×10^{-10}N nm/rad², respectively [202, 206, 207]. Substitution of these values into Eqs (3.50)–(3.51) results in a Young's modulus of 5.488×10^{-8} N/A°², shear modulus of 8.701×10^{-9} N/A°² and a diameter (d_b) of 1.466A° for the structural beam element. These parameters are fed as inputs to the FE model. Also, the compressive-force displacement relationship of these spring elements can be approximated through the Lennard-Jones force, see Eq. (49). Assuming displacement to be the change in interatomic distance relative to the critical distance of 0.38 nm (at $r = 0.38$ nm, the Lennard-Jones force is zero [37, 208], the non-linear stiffness-displacement relationship of the spring elements is obtained and shown in Fig. 3.7. The FE models developed for different configurations of nanotubes are illustrated in Fig. 3.8c corresponds to the FE model for a typical double-walled nanotubes set.

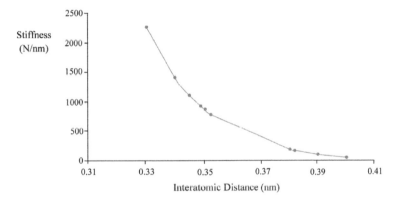

FIGURE 3.7 Variation of the structural spring with inters atomic distance.

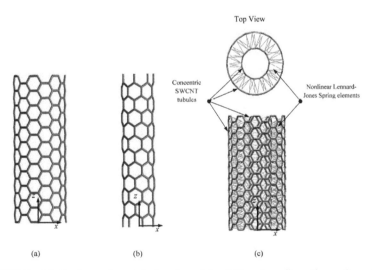

FIGURE 3.8 Finite element models developed for different configurations of nanotubes. (a) Armchair (7, 7), (b) zig-zag (7, 0) and (c) double-wall nanotubes (5, 5) and (10, 10).

The finite element model generation of double- or multi-walled nanotubes can be realized as the generation of two concentric tubules and subsequent creation of the spring elements between nodes of the concentric tubules when the distance between them is less than 3.8A° but greater than 3A°.

Subsequent to their creation, the FE models are investigated for their elastic response under the action of pure tension and pure torsion. Simulation of pure tension was achieved by constraining both translational and rotational degrees of freedom on the nodes at one edge (lower end) of the nanotubes, while the other end was loaded in pure tension. Likewise, the condition of pure torsion on nanotubes is obtained by completely constraining the nodes at one edge (lower end), while the nodes at the upper end were constrained from moving in radial direction (UR = 0). A uniform tangential force was then applied on all nodes at the upper end. These considerations are illustrated in Figs. 3.9 and 3.10.

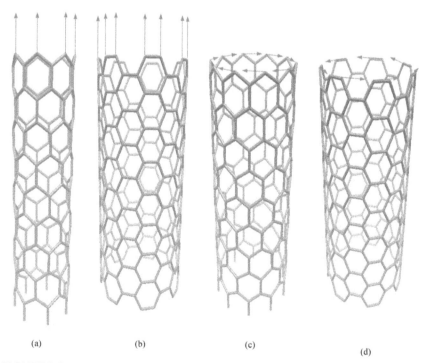

(a)   (b)   (c)   (d)

FIGURE 3.9   Boundary and loading conditions on the FE models of typical single-walled nanotubes (a) Zig-zag (7, 0), (b) Armchair (7, 7), (c) zig-zag (10, 0) and (d) armchair (7, 7).

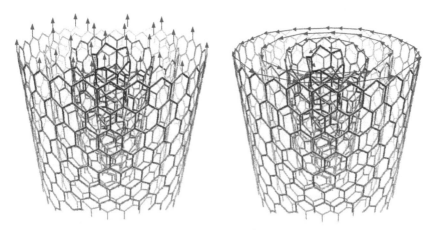

FIGURE 3.10 Boundary and Loading conditions on the FE models of typical multi-walled nanotubes sets (a) Four walled zig-zag nanotubes set (5, 0)(14, 0)(23, 0)(32, 0) under pure tension and (b) four walled armchair (7, 7), (c) zig-zag (10, 0) and (d) armchair (7, 7).

3.4.6 MODELING RESULTS AND DISCUSSION

3.4.6.1 RESULTS FROM THE ASYMPTOTIC HOMOGENIZATION TECHNIQUE

The elastic properties of a homogenized SWCNT can be obtained from Eq. (29) as follows [197]:

$$\begin{bmatrix} N_{11} \\ N_{22} \\ N_{12} \end{bmatrix} = \delta^2 \frac{E}{l} \frac{\pi}{16\sqrt{3}} \begin{bmatrix} 3 & 1 & 0 \\ 1 & 3 & 0 \\ 0 & 0 & 1 \end{bmatrix} \begin{bmatrix} \varepsilon_{11} \\ \varepsilon_{22} \\ \varepsilon_{12} \end{bmatrix} \tag{52}$$

The elastic modulus in the α_1 direction can be calculated by assuming that only load N_{11} acts on the structure, i.e., $N_{22} = N_{12} = 0$. Using these values in Eq. (3.33) we get

$$\begin{bmatrix} \varepsilon_{11} \\ \varepsilon_{22} \\ \varepsilon_{12} \end{bmatrix} = \frac{1}{E\delta^2} \frac{16\sqrt{3}}{\pi} \begin{bmatrix} 3 & 1 & 0 \\ 1 & 3 & 0 \\ 0 & 0 & 1 \end{bmatrix}^{-1} \begin{bmatrix} N_{11} \\ 0 \\ 0 \end{bmatrix} \tag{53}$$

The effective modulus in the α_1 direction is given as

$$E_{11} = \frac{\sigma_{11}}{\varepsilon_{11}} = \frac{N_{11} / \delta}{\varepsilon_{11}} = \frac{\pi}{6\sqrt{3}} \left(\frac{\delta E}{l} \right) \tag{54}$$

Similarly, the effective modulus in the direction of α_2 can be calculated by applying N_{22} alone, to give

$$E_{22} = \frac{\sigma_{22}}{\varepsilon_{22}} = \frac{N_{22} / \delta}{\varepsilon_{22}} = \frac{\pi}{6\sqrt{3}} \left(\frac{\delta E}{l} \right) \tag{55}$$

In Eqs (52)–(55) and in the sequel, by r and e we denote stresses and strains, not to be confused with the Lennard-Jones parameters in Eqs (44) and (49). The effective shear modulus G_{12} can be obtained by applying only N_{12}. The result is [197]:

$$G_{12} = \frac{\tau_{12}}{\varepsilon_{12}} = \frac{N_{12} / \delta}{2\varepsilon_{12}} = \frac{\pi}{32\sqrt{3}} \left(\frac{\delta E}{l} \right) \tag{56}$$

It can thus be observed that the effective modulus of the structure in both α_1 and α_2 directions is the same. Therefore, the Young's modulus of the SWCNT along the longitudinal direction can be given as [197]:

$$E_{SWCNT} = \frac{\pi}{6\sqrt{3}} \left(\frac{\delta E}{l} \right) \tag{57}$$

Equations (56) and (57) represent explicit relationships for the moduli of SWCNT in terms of material and geometric parameters. As an illustration, for the purpose of an effective comparison with the finite element method, let us use the expressions for k_r, k_θ and k_φ (for a circular cross-section) in Eq. (48) and the typical values for these force constants:

$$E = 5.488 \times 10^{-6} \, N / nm^2; \delta = 0.147 nm; L = 0.142 nm \tag{58}$$

\Substituting these quantities in Eq. (57) yields:

$$E_{SWCNT} = 1.71 \, TPa \tag{59a}$$

The corresponding shear modulus can be calculated using Eq. (58), and is

$$G_{SWCNT} = 0.32 \, TPa \tag{59b}$$

In the next section, corresponding values based on the FE approach are presented.

3.4.6.2 RESULTS FROM THE NUMERICAL FE BASED METHOD

To calculate the effective properties of a nano tube based on the FE approach it is assumed that the nanotubes can be represented by an equivalent continuum tube shown in Fig. 3.12. The Young's modulus is then given as

$$E_{NT} = \frac{\sigma}{\varepsilon} = \frac{P / A_0}{\Delta L / L_0} \tag{60}$$

Here, P is the total axial force applied, A_0 is the cross-sectional area of the equivalent continuum tube, ΔL is the corresponding axial deformation obtained from the finite element model and L is the length of the nanotubes under investigation.

Similarly, the shear modulus of the carbon nanotubes is:

$$G_{NT} = \frac{TL_0}{0 J_0} \tag{61}$$

where T is the total torque applied on the nanotubes, θ is the angle of twist (obtained from finite element solution) and J_0 is the polar moment of inertia for the continuum tube model of nanotubes. The parameters A_0 and J_0 for single-walled or multi-walled carbon nanotubes can be calculated by considering the cross-section of the continuum models as shown in Fig. 3.11. For single-walled nanotubes of mean radius R_{NT}, the cross-sectional area and polar moment of inertia can be obtained as follows:

$$A_0 = \pi \left[(R_{NT} + \frac{t}{2})^2 - (R_{NT} - \frac{t}{2})^2 \right] \tag{62a}$$

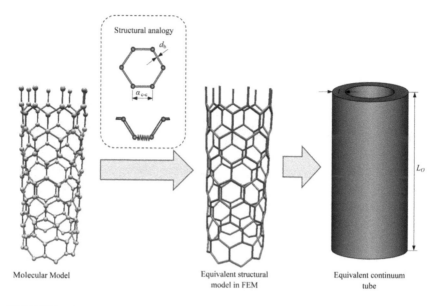

FIGURE 3.11   Equivalence of molecular finite element and continuum model.

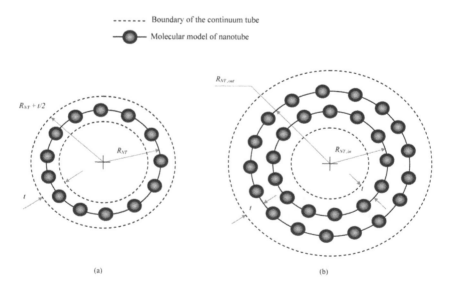

FIGURE 3.12.   Cross-sectional equivalence of nanotubes and their continuum tube models (a) Singled walled carbon nanotubes and (b) multi walled carbon nanotubes.

$$J_0 = \frac{\pi}{2}\left[(R_{NT} + \frac{t}{2})^4 - (R_{NT} - \frac{t}{2})^4\right]$$ (62b)

Similarly if RNT, in and RNT, out are the inner and outer radii of a multi-walled carbon nanotubes, then the cross sectional area and polar moment of inertia are given by

$$A_0 = \pi\left[\left(R_{NT,out} + t\right)^2 - \left(R_{NT,in} - t\right)^2\right]$$ (63a)

$$J_0 = \frac{\pi}{2}\left[\left(R_{NT,out} + t\right)^4 - \left(R_{NT,in} - t\right)^4\right]$$ (63b)

The radii R_{NT}, $R_{NT,in}$ and $R_{NT,out}$ are based on the corresponding chiral indices. Based on the formulation described above, Young's and shear moduli are calculated for both single- and double-walled carbon nanotubes. Both zig-zag and armchair configurations are investigated and compared. The carbon–carbon bond length (a_{c-c}) was assumed to be 0.142 nm. However, no established values are available for the wall thickness, t, of the nanotubes. The values of the wall thickness as suggested by available literature, varied significantly from 0.066 nm [12, 186] to 0.34 nm [47, 209] and 0.68 nm [210]. In view of such a wide range of suggested wall thickness, a study is performed using the FE model presented herein. Results indicate that an appropriate wall thickness for the proposed modeling approach is about 0.68 nm. This value of wall thickness is in excellent agreement with the work of Odegard et al. [210, 211].

3.5 STRUCTURAL MECHANICS APPROACH TO CARBON NANOTUBES

From the structural characteristics of carbon nanotubes, it is logical to anticipate that there are potential relations between the deformations of carbon nanotubes and frame-like structures. For macroscopic space frame structures made of practical engineering materials, the material properties and element sectional parameters can be easily obtained from material data handbooks and calculations based on the element sectional dimensions. For nanoscopic carbon nanotubes, there is no information about the elastic and sectional properties of the carbon-carbon bonds and the material properties. Therefore, it is imperative to establish a linkage between the microscopic computational chemistry and the macroscopic structural mechanics.

3.5.1 POTENTIAL FUNCTIONS OF MOLECULAR MECHANICS

From the viewpoint of molecular mechanics, a carbon nano tube can be regarded as a large molecule consisting of carbon atoms. The atomic nuclei can be regarded as material points. Their motions are regulated by a force field, which is generated by electron-nucleus interactions and nucleus-nucleus interactions [212]. Usually, the force field is expressed in the form of steric potential energy. It depends solely on the relative positions of the nuclei constituting the molecule (Fig. 3.13).

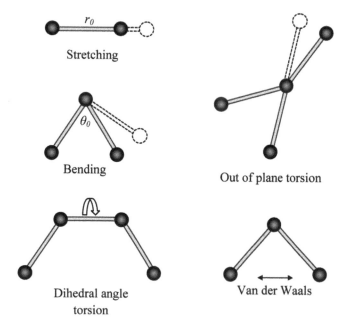

FIGURE 3.13   Inter atomic interactions in molecular mechanics.

The general expression of the total steric potential energy, omitting the electrostatic interaction, is a sum of energies due to valence or bonded interactions and non bonded interactions [213]:

$$U = \sum U_r + \sum U_\theta + \sum U_\phi + \sum U_w + \sum U_{vdw}, \tag{64}$$

where U_r is for a bond stretch interaction, U_θ for a bond angle bending, U_ϕ for a dihedral angle torsion, U_ω for an improper (out-of-plane) torsion, U_{vdw} for a non bonded van der Waals interaction, as shown in Fig. 3.14. There has been a wealth of literature in molecular mechanics devoted to finding the reasonable functional forms of these potential energy terms [201, 202, 213, 214]. Therefore, various functional forms may be used for these energy terms, depending on the particular material and loading conditions considered. In general, for covalent systems, the main contributions to the total steric energy come from the first four terms, which have included four-body potentials. Under the assumption of small deformation, the harmonic approximation is adequate for describing the energy [203]. For sake of simplicity and convenience, we adopt the simplest harmonic forms and merge the dihedral angle torsion and the improper torsion into a single equivalent term, that is,

$$U_r = \frac{1}{2} k_r (r - r_0)^2 = \frac{1}{2} k_r (\Delta r)^2, \tag{65}$$

$$U_\theta = \frac{1}{2} k_\theta (\theta - \theta_0)^2 = \frac{1}{2} k_\theta (\Delta \theta)^2, \tag{66}$$

$$U_t = U_\phi + U_w = \frac{1}{2} k_\tau (\Delta \phi)^2, \tag{67}$$

where k_r, k_θ and k_τ are the bond stretching force constant, bond angle bending force constant and Torsional resistance, respectively, and the symbols Δr, $\Delta \theta$ and $\Delta \phi$ represent the bond stretching increment, the bond angle change and the angle change of bond twisting, respectively.

3.5.2 LINKAGE BETWEEN SECTIONAL STIFFNESS PARAMETERS AND CONSTANTS OF FORCE FIELDS

In a carbon nano tube, the carbon atoms are bonded to each other by covalent bonds and form hexagons on the wall of the tube. These covalent bonds have their characteristic bond lengths and bond angles in a three-dimensional space. When a nano tube is subjected to external forces, the displacements of individual atoms are constrained by these bonds. The total deformation of the nanotubes is the result of these bond interactions. By considering the covalent bonds as connecting elements between carbon atoms, a nano tube could be simulated as a space frame-like structure. The carbon atoms act as joints of the connecting elements.

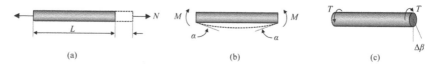

(a)   (b)   (c)

FIGURE 3.14   Pure tension, bending and torsion of an element.

In the following, the relations between the sectional stiffness parameters in structural mechanics and the force constants in molecular mechanics are established. For convenience, the sections of carbon-carbon bonds are assumed to be identical and uniformly round. Thus it can be assumed that $I_x = I_y = I$ and only three stiffness parameters, EA, EI and GJ, need to be determined.

Because the deformation of a space frame results in the changes of strain energies, the three stiffness parameters based on the energy equivalence are determined. Notice that each of the energy terms in molecular mechanics (the Eqs. (65)–(67))represents an individual interaction and no cross-interactions are included, the strain energies of structural elements under individual forces needs to be considered, too. According to the theory of classical structural mechanics, the strain energy of a uniform beam of length L subjected to pure axial force N (Fig. 3.14a) is

$$U_A = \frac{1}{2}\int_0^L \frac{N^2}{EA}\,dL = \frac{1}{2}\frac{N^2 L}{EA} = \frac{1}{2}\frac{EA}{L}(\Delta L)^2,\qquad(68)$$

where ΔL is the axial stretching deformation. The strain energy of a uniform beam under pure bending moment M (Fig. 3.14b) is

$$U_M = \frac{1}{2}\int_0^L \frac{M^2}{EI}\,dL = \frac{2EI}{L}\alpha^2 = \frac{1}{2}\frac{EI}{L}(2\alpha)^2,\qquad(69)$$

Where α denotes the rotational angle at the ends of the beam. The strain energy of a uniform beam under pure torsion T (Fig. 3.14c) is

$$U_T = \frac{1}{2}\int_0^L \frac{T^2}{GJ}\, dL = \frac{1}{2}\frac{T^2 L}{GJ} = \frac{1}{2}\frac{GJ}{L}(\Delta\beta)^2,$$ (70)

where $\Delta\beta$ is the relative rotation between the ends of the beam.

It can be seen that in Eqs. (65)–(70)both U_r and U_A represent the stretching energy, both U_θ and U_M represent the bending energy, andboth U_τ and U_T represent the Torsional energy. It is reasonable to assume that the rotation angle 2α is equivalent to the total change $\Delta\theta$ of the bond angle, ΔL is equivalent to Δr, and $\Delta\beta$ is equivalent to $\Delta\varphi$. Thus by comparing Eqs (65)–(67) and Eqs (68)–(70), a direct relationship between the structural mechanics parameters EA, EI and GJ and the molecular mechanics parameters k_r, k_θ and k_τ is deduced as following:

$$\frac{EA}{L} = k_r, \frac{EI}{L} = k_\theta, \frac{GJ}{L} = k_\tau.$$ (71)

The Eq. (71) establishes the foundation of applying the theory of structural mechanics to the modeling of carbon nanotubes or other similar fullerene structures. As long as the force constants k_r, k_θ and k_τ are known, the sectional stiffness parameters EA, EI and GJ can be readily obtained. And then by following the solution procedure of stiffness matrix method for frame structures, the deformation and related elastic behavior of graphene sheets and nanotubes at the atomistic scale can be simulated (Fig. 15).

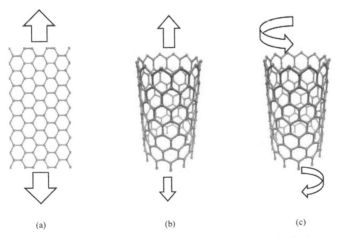

(a)   (b)   (c)

FIGURE 3.15   Computational models of a graphene sheet and single-walled carbon nanotubes.

To verify the reliability and efficiency of the structural mechanics approach to the modeling of carbon nanotubes and to demonstrate its capability, the graphite sheets and single-walled carbon nanotubes are chosen as examples and calculate some of their basic elastic properties, such as Young's modulus and shear modulus. In these computations, the initial carbon-carbon bond length is taken as 1.421A° [8]. The computational results are compared with the limited existing theoretical and experimental results.

3.6   YOUNG'S MODULUS OF A GRAPHENE SHEET

As stated earlier, a carbon nano tube can be viewed as a sheet of graphite that has been rolled into a tube. Thus, the Young's modulus of a graphene sheet is calculated (Fig. 3.14a) to verify the feasibility of the present method. It is also expected that these calculations can provide useful information concerning the selection of force field constants. Young's modulus of a material is the ratio of the normal stress to the normal strain in a uniaxial tension test.

$$Y = \frac{\sigma}{\varepsilon} = \frac{F/A_0}{\Delta H/H_0},$$

(72)

where F stands for the total force acting on the atoms at one end of the sheet, $A_0 = Wt$ is the cross-sectional area of the sheet with width W and thickness t, H_0 is the initial length and ΔH its elongation. The thickness t is taken as the interlayer spacing of graphite, 0.34 nm. [8]. Table 3.1 lists the computed Young's modulus of a graphene sheet for different model sizes and different force field parameters, which are selected from AMBER [8, 202]. It can be seen that our results from both pairs of force field constants are fairly close to the commonly accepted value (1.025 TPa) [215] of the Young's modulus of graphite. Furthermore, our results are very close to the recent prediction (1.029 TPa)based on Ab initio computation [216]. It is also observed in Table 3.1 that the computational results are weakly affected by model size.

3.6.1 YOUNG'S MODULUS OF A SINGLE-WALLED CARBON NANO TUBE

The successful prediction of the Young's moduli of graphene sheets established our confidence in applying this method to analyzing the Young's modulus of carbon nanotubes. Two main types of carbon nanotubes, that is, armchair and zigzag, are considered. The force constant values are chosen basedupon the experience with graphite sheets: $k_r/2$= 469 kcal mol^{-1}A$^{°-2}$, $k_\theta/2$ = 63 kcal mol^{-1} rad^{-2}. The force constant k_τ is adopted as $k_\tau/2$= 20 kcal mol^{-1} rad^{-2}based on references [202, 217]. There is no test data for optimizing the choice of k's. But our calculation showed that the influence of k's on carbon nanotubes Young's modulus is very weak. The tensile force is applied on one end of a carbon nanotubes and the other end is fixed (Fig. 3.14b). The Eq. (72) is still used for calculating the Young's modulus, except that the sectional area is now A_0= πdt with d standing for the tube diameter.

Figure 3.15 displays the variations of the Young's modulus with nanotubes diameter. It can be seen that the trend is similar for both armchair and zigzag SWNTs, and the effect of nanotubes chirality is not significant. For tube diameters larger than 0.7 nm, the Young's moduli of zigzag nanotubes become slightly larger than those of armchair nanotubes, and the trend is reversed for tube diameters less than 0.7 nm. This correla-

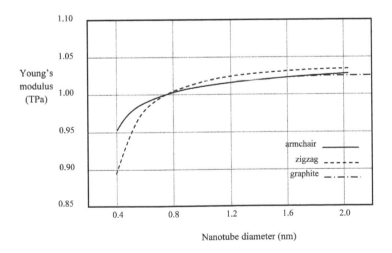

FIGURE 3.15 Young's moduli of carbon nanotubes versus tube diameter.

tion between nanotubes chirality and Young's modulus can be understood from inter atomic bond orientation. In zigzag nanotubes, one-third of the bonds are aligned with the loading direction, while every bond is at an angle with the loading axis in armchair nanotubes. Thus the difference in nanotubes elastic behavior is a direct consequence of the intrinsic atomic structure. When the tube diameter is very smaller, the distortion of C–C bonds of zigzag nanotubes may be more significant than that of armchair nanotubes. Therefore, the Young's modulus of the former is slightly less than that of the latter.

3.6.2. SHEAR MODULUS OF A SINGLE-WALLED CARBON NANO TUBE

Due to the difficulty in experimental techniques, there is still no report on the measured values of shear modulus of carbon nanotubes. Theoretical predictions on the shear modulus of carbon nanotubes are also very few. Lu [47] predicted the shear modulus for carbon SWNTs by using empirical lattice dynamics model, and concluded that the shear modulus (0.5 TPa) is comparable to that of diamond and is insensitive to tube diameter and tube chirality. Popov et al. [32] also used the lattice-dynamics model and derived an analytical expression for the shear modulus. Their results indicated that the shear moduli of carbon SWNTs are about equal to that of graphite for large radii but are less than that of graphite at small radii, and the tube chirality has some effect on the shear modulus for small tube radii.

Here, a carbon SWNT is assumed to be subjected to a Torsional moment at one end and is constrained at the other end (Fig. 3.17). The following formula, which is based on the theory of elasticity at the macroscopic scale, is used for obtaining the shear modulus:

$$S = \frac{TL_0}{\theta J_0}$$

(73)

where T stands for the torque acting at the end of an SWNT, L_0 is the length of the tube, θ is the Torsional angle and J_0 for the cross-sectional polar inertia of the SWNT. In the calculation of the polar inertia J_0, the SWNT is treated as a hollow tube with a wall thickness of 0.34 nm.

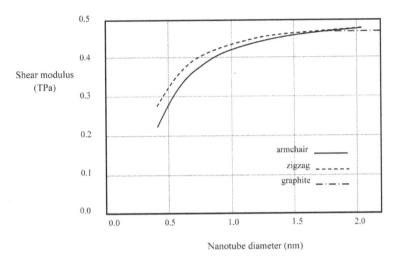

FIGURE 3.17   Shear moduli of carbon nanotubes versus tube diameter.

Figure 3.17 illustrates the computational results for armchair and zig-zag SWNTs. It is observed that the shear modulus behaves similarly to the Young's modulus in that it increases with increasing tube diameter for small tube diameters. At larger diameters (e.g., >2 nm), the shear modulus becomes insensitive to tube diameter. For the range of tube diameter considered, the effect of tube chirality on the shear modulus is not significant. For tube diameters larger than 2.0 nm, the shear moduli of armchair and zigzag SWNTs are almost the same. By comparing our findings with existing results [32, 47], it can be concluded that the present method can achieve the same accuracy as other existing methods.

KEYWORDS

- **Cylindrical Modeling**
- **Finite Element**
- **Molecular Mechanics**
- **Nanotubes**
- **Shear Modulus**
- **Young's Modulus**

CHAPTER 4

COMPUTATIONAL MECHANICS MODELING

CONTENTS

4.1   INTRODUCTION

CNTs are at the boundary between small structures and large molecules. Reflecting their condition, single-scale modeling approaches can be generally classified into two categories: atomistic and continuous modeling. For the former approach, the methods are very general [218] but their operable application is limited to small and short-lived systems because they are computationally expensive. In the past, interactions were obtained by two-body potentials with simple analytical form (Morse and Lennard-Jones). Recently, many-body terms are accounted in the most accurate potentials, as described in the following sections. As a result, their complexity is usually polynomial but seldom linear on the number of atoms [219]. The latter approach simplifies the model by treating CNTs as continuous elastic shells, extending the already large and fruitful knowledge from CM. However, many of these models seems to be far too reductionist. To some extent, their basic approaches neglect the detailed characteristics of chirality and are almost unable to account for forces acting on individual atoms (and their eventual vacancies that constitute realistic defects). Many authors have analyzed the mechanics of carbon nanotubes through CM, but the relevance of continuous models for systems composed by few atoms is controversial [40, 48, 220–222].

Traditional approaches to modeling nanostructures in the framework of CM are based on the existing theories, like Kirchhoff elastic rod theory [223] and thin shell theory [12, 40, 224]. In these approaches, the parameters of the theoretical setting are fitted from experimental data, which empirically come from the discrete model. However, SWCNTs can be appropriately described as objects of reduced dimensionality in the three-dimensional Euclidean space, since graphene is a single-layer atomic film. This comes from the nature of the rolled graphene lattice, which is basically a two-dimensional arrangement of atoms, and whose energy, according to the Born-Oppenheimer approximation (BOA), exclusively depends on the positions of the atoms and doesn't suggest any meaningful thickness. In both discrete and continuous modeling a number of specialized theories have been developed or adapted for CNTs. In the former sections, some of the most influential and useful approaches are described. This will lead to the combined approach, which can employ in a coherent framework some of the advantages of the single-scale methods.

4.2 MOLECULAR MECHANICS

The behavior and the energetic of molecules are fundamentally quantum mechanical. To some extent, the mechanical aspects of a molecular system can be modeled with the use of Newtonian Mechanics (NM) that can approximate some results of Quantum Mechanics (QM). In these formalizations, the potential energy of the molecular system is obtained by the computation of an appropriate force field (FF). In this context, a FF is a functional form coupled with a set of parameters that describe the potential energy of the molecular system.

Here, the term "FF" refers to the functional form with the parameter sets used to describe the potential energy of a simulated system of particles. The term FF in computational chemistry and biology has a different usage from the standard one in physics: it denotes a potential function instead of the negative gradient of a scalar potential. Molecular systems are modeled as systems of particles and often the particles are identified with the atoms of the structure.

In some significant cases, the system's energy is expressed only as a function of the nuclear positions, coherently with the BOA. Being an atomistic method, every atom that constitutes the molecular system is modeled as a single material particle, equipped with mass ma and Van der Waals radius r_a. The formulation of the system's energy allows the prediction of the relative energies between different conformations or between different molecules. Also, it permits the identification of the equilibrium geometries and of the transition states, which correspond to potential energy surface local minima and first-order saddle points, respectively. Molecular Mechanics (MM) is a family of methods that uses Taylor and Fourier series expansions and additional terms, all of which involve empirically fitted parameters.

The range of applicability of MM is broad and well known; its great efficiency is vastly employed in computational chemistry, material science and molecular biology. This is motivated by the fact that its computational cost in memory and time is probably the lowest of any detailed computational chemistry method. From small molecules to large biological systems, MM is a refinement tool for structure determination, free energy prediction and a model for molecular motion. A specific potential is introduced in order to predict the energy associated with the given conformation of the molecule. It is important to point out that it doesn't absolutely quantify the energy of the system: only differences in energy between conformations are physically meaningful. MM can be used to study small molecules as well as large biological systems or material assemblies with many thousands to millions of atoms. The methods which minimize the potential energy are known as energy minimization techniques and the most used are the Steepest Descent (SD) [225, 226] and the Conjugate Gradient (CG) [226–228] methods.

4.3 PRINCIPLES AND ENERGY FORMULATIONS

Schematically, MM is based on the following principles:

- particles are spherical and possess a net charge;
- particles can contain agglomerates of atoms or single atoms, where nuclei and electrons are included in the same particle;
- interactions are pre assigned to specific sets or subsets of atoms;
- interactions are based on elastic springs and classical potentials;
- interactions determine the spatial distribution of the particles and of their energies.

A basic model in MM considers particles as spheres and bonds as springs. The mathematics of spring deformation is used to describe the ability of bonds to twist, bend and stretch. Basic MM considers only a static bonding of the atoms, while some more refined extensions may include a dynamical reconfiguration.

In both the situations, non-bonded atoms interact through van der Waals attraction, electrostatic interactions and steric repulsion. These properties are the easiest to describe mathematically when the particles are considered as spheres with their characteristic radii.

The goal of MM is to predict the energy associated with a given conformation of a molecule. This is done by the definition of a molecular configuration $\psi \in R^n$ and of an energy function E: $R^n \rightarrow$ R. In the simplest case, the rotational Vibrational energy levels of a diatomic with the configuration are given by:

$$E\left(\psi_e, v_e, \mu_e, B_e, D_e, \alpha_e, Y_{00}\right) \cong U\left(\psi_e\right) + hv_e\left(v + \frac{1}{2}\right) - h\mu_e(v + \frac{1}{2})^2 + hB_e J\left(J+1\right)$$

$$-h\alpha_e\left(v + \frac{1}{2}\right)J\left(J+1\right) - hD_e J^2\left(J+1\right)^2 + Y_{00}. \tag{1}$$

This is a formulation where E is a function of the spectroscopic constants

$$\psi_e, v_e, \mu_e, B_e, \overline{D}_e, \alpha_e, Y_{00}$$

and of the quantum numbers

$$v, J.$$

By definition, $\mu_e = v_e x_e$ and ψ_e is given by the configuration at the equilibrium.

Then the high-resolution spectra can be fit to such expressions, in order derive the spectroscopic constants from the experimental data. In MM, the analysis starts with the formulation of a classical and simpler expression for the energy E as a function of the nuclear coordinates, even if the atoms can possess given initial velocities. Then, it applies this to ground states only, not directly accounting for Vibrational, rotational or electronic excitations. In this classical representation, no special quantized energy levels v and J are present and it's possible to express the Taylor-series expansion of the potential energy as a function of the nuclear coordinate's ψ. Again in the simplest case, for a diatomic molecule one has $\psi = \psi$ and gets the expansion

$$E(\psi) = U(\psi_e) + \sum_{j=1}^{+\infty} \frac{1}{j!} \frac{d^j E(\psi)}{d\psi j}(\psi - \psi_e)^j. \tag{2}$$

The second derivative at $\psi = \psi_e$ is identified as the harmonic oscillator component and the third derivative is related to the harmonic constant $\omega_e x_e$, which includes also a term of the fourth derivative. Therefore, the derivatives can be treated as parameters that are identified by experimental outcomes, producing the convenient form

$$E(\psi) = \sum_{j=2}^{n} k_j (\psi - \psi_e)^j, \tag{3}$$

that approximates the Taylor expansion with coefficients going from 2 to n, absorbing in those terms the constants $1/(n!)$. In Eq. (3), the term $U(\psi_e)$ doesn't appear, because the arbitrary zero of the energy at the equilibrium has been set to the zero value. Also, the first term of the sum in Eq. (2) dropped out, since the expansion is around the equilibrium configuration ψ_e, where the gradient is zero.

The central idea in MM is that the constants obtained from experiments on the simplest diatomic model can be transferred to other, more complex, molecules. Indeed, bond lengths are almost the same in every molecule and the same applies for the stretching frequencies. This approach is improved by specializations as a function of the bond type and of the peculiar conditions of the studied object. For instance, C-H bond lengths range between 1.06 and 1.10°A in almost every molecule at room conditions, while the stretching frequencies are included between the bounds at 2900 and at 3300 cm^{-1}. As a result, a C-H bond has a similar ψ_e and v_e for most of the molecules. Refinements permit to separate sp2 by sp3 carbons and, in the case of a graphene sheet, one has that $\psi_e = a_{(C-C)}$

For more complex molecules, the total energy is expressed as a sum of Taylor series expansions for stretches of every pair of bonded atoms and adds supplementary potential energy terms coming from Torsional energy, bending, van der Waals energy, and electrostatics and cross terms. The result is the following separation of the components:

$$E(\psi) \simeq E_s(\psi) + E_b(\psi) + E_t(\psi) + E_c(\psi) + E_v(\psi) + E_e(\psi). \qquad (4)$$

Decomposition of E into the sum of energy components is justified by the fact these functions concentrates on several separate aspects. Precisely,

$$E_s : R^n \to R$$

Accounts for the global stretching energy, E_b: $R^n \to R$ for the global bending energy, E_t: $R^n \to R$ for the global Torsional energy, E_b: $R^c \to R$ for the energy of the cross terms and, finally, E_n: $R^n \to R$ accounts for the all the energy of the non-bonded interactions. By separating the van der Waals E_v and the electrostatic E_e terms and aggregating them into E_n, MM attempts to make the remaining constants more transferable among molecules than they would are in a spectroscopic FF. Many different kinds of FFs have been described and rapidly matured over the years. Some formulations include additional energy terms that describe other types of deformations. Some other FFs account for the coupling between bending and stretching of adjacent bonds, in order to improve the accuracy of the mechanical model. The number of extensions to the basic formulation has lead to a large variety of specializations to limited areas of interest.

Stretching Energy Whenever a bond is compressed or stretched, the energy E_s grows. The stretching action is visualized by Fig. 4.1. Let ψ_{12} be the actual distance between the positions of the atoms a_1 and a_2 and let ψ_0 represent the equilibrium distance of the bond. Then the energy potential for bond stretching and compressing is described by the Taylor series

$$E(\psi_{12}) = \sum_{i=2}^{+\infty} k_j (\psi_{12} - \psi_0)^j \simeq \sum_{i=2}^{n} k_j (\psi_{12} - \psi_0)^j. \tag{5}$$

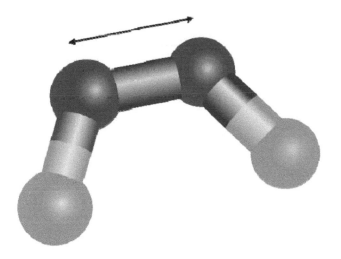

FIGURE 4.1   Representation of a stretching action.

Of course, all the FFs truncate this series but they retain different numbers of terms in this expansion. The most straightforward truncation is the application to the Hooke's law with $n = 2$, which yields the harmonic approximation. It's immediate to notice that a diatomic molecule which is actually bound using a harmonic potential can never dissociate.

Not surprisingly, when more terms are included, alternative expansions exhibit an incorrect limiting behavior and can only is applied when the local deformation is very small. When the distances are large, the highest powers of $(\psi_{12} - \psi_0)$ dominate, leading $E(\psi_{12})$ to a positive or negative explosion, depending on the sign and the magnitude of kn, kn^{-1}.

Consequently, several solutions have been proposed: an extension of the Hooke's law for a spring with the addition of a cubic term and the Morse potential [229]. In the former method, the energy has the form

$$E_{s3}\left(\psi_{12}\right)=k_{2}\left(\psi_{12}-\psi_{0}\right)^{2}\left(1-2\left(\psi_{12}-\psi_{0}\right)\right) \tag{6}$$

And the cubic term helps to keep the energy from rising too sharply as the bond is stretched, as shown by its normalized plot in Fig. 4.2. In the latter method, the Morse potential is a simple function with a correct limiting behavior that has the form

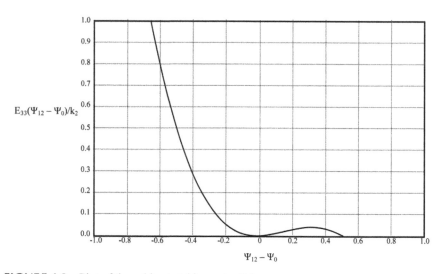

$E_{33}(\Psi_{12}-\Psi_{0})/k_2$

$\Psi_{12}-\Psi_0$

FIGURE 4.2   Plot of the cubic stretching potential.

$$E_{sM}\left(\psi_{12}\right)=D(1-e^{-\rho\left(\psi_{12}-\psi_{0}\right)})^{2}, \tag{7}$$

where D is the dissociation energy (that is the well depth) and

$$\rho=\sqrt{k/2D},$$

which controls the width of the potential. Also, the constant k is the force at the minimum of the well. Figure 4.3 plots the normalized Morse potential and it's $L^2([a, b])$ harmonic best approximation in the interval given by

$$-\frac{1}{2} \le \rho\left(\psi_{12} - \psi_0\right) = y \le \frac{1}{2}. \tag{8}$$

In a generic interval [a, b] of R, the $L^2([a, b])$ best approximation finds $k_2 \in R$ such that the error

$$e_{ab}\left(k_2\right) = \sqrt{\int_a^b (1 - e^{-y})^2 - k_2 y^2)^2 \, dy} \tag{9}$$

is minimized; that is,

$$k_2 = \arg\left[\min_{k \in R} \min_{k \in R} \left\{e_{ab}(k_2)\right\}\right] = \arg\left[\min_{k \in R} \min_{k \in R} \left\{\sqrt{\int_a^b \left((1 - e^{-y})^2 - k_2 y^2\right)^2 dy\right)}\right\}\right]$$

must holds. The problem is well posed, and the equivalent form is

$$k_2 \in R : \frac{d}{dk} \sqrt{\int_a^b \left((1 - e^{-y})^2 - k_2 y^2\right)^2 dy} = 0 \tag{10}$$

Since the solution exists and it's unique.

In fact, the problem of best approximation in $L^2([a, b])$ of the normalized Morse potential

$$E_M\left(y\right) = (1 - e^{-y})^2 \tag{11}$$

in the real interval [a, b] by an harmonic potential

$$E_h\left(y\right) = k_2 y^2 \tag{12}$$

with $k_2 \in R$, has one and only one solution. The best approximation problem has the formulation that follows: find $k_2 \in R$ such that the error of Eq. (9) is minimized. One has that

$$f(k_2) = \int_a^b (1 - e^{-y})^2 - k_2 y^2)^2 \, dy = \frac{e^{-4b}}{60}(12b^5 e^{4b} k_2^2 + (-40b^3 e^{4b} + (-240b^2 - 480b - 480)e^{3b} - 15 + (60b^2 + 60b + 30)e^{2b})k_2 + 80e^b + 60be^{4b} + 240e^{3b} - 180e^{2b}) - \frac{e^{-4a}}{60}(12a^5 e^{4a} k_2^2 + (-40a^3 e^{4a} + (-240a^2 - 480a - 480)e^{3a} - 15 + (60a^2 + 60a + 30)e^{2a})k_2 + 80e^a + 60ae^{4a} + 240e^{3a} - 180e^{2a}) = \alpha_0 + \alpha_1 k_2 + \alpha_2 k_2^2 \tag{13}$$

and that

$$\forall k_2 \in Rf\left(k_2\right) = \alpha_0 + \alpha_1 k_2 + \alpha_2 > 0, \tag{14}$$

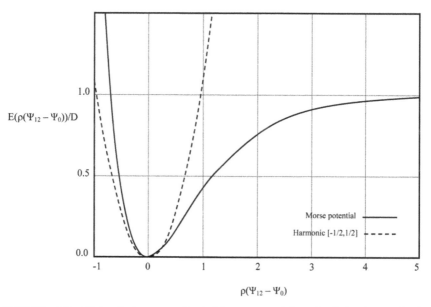

FIGURE 4.3 Plot of the normalized Morse potential and of its harmonic best approximation in $[-\frac{1}{2}, \frac{1}{2}]$.

so the same k_2 will also minimize the error function

$$e_{ab}(k_2) = \sqrt{f(k_2)}$$

Consequently, it's possible to obtain the first derivative of $f(k_2)$, that is

$$\frac{df(k_2)}{dk_2} = \frac{d}{dk_2}\int_a^b ((1-e^{-y})^2 - k_2 y^2)^2 dy == \frac{e^{-4b}}{60}\left(24b^5 e^{4b}k_2(-40b^3 e^{4b} + (-240b^2 - 480b - 480)e^{3b} + (60b^2 + 60b + 30)e^{2b})\right) - \frac{e^{-4a}}{60}\left(24a^5 e^{4a}k_2(-40a^3 e^{4a} + (-240a^2 - 480a - 480)e^{3a} + (60a^2 + 60a + 30)e^{2a})\right) = 0 \quad (15)$$

that has one zero, namely

$$k_2 = \frac{20b^3 e^{2b+2a} - 20a^3 e^{2b+2a} - 120a^2 e^{2b+a} - 240ae^{2b+a} - 240e^{2b+a}}{(12b^5 e^{2b+2a} - 12a^5 e^{2b+2a})} +$$
$$\frac{b(240e^{b+2a} - 30e^{2a}) + b(120e^{b+2a} - 30e^{2a})}{(12b^5 e^{2b+2a} - 12a^5 e^{2b+2a})} + \frac{240e^{b+2a} + (30a^2 + 30a + 15)e^{2b} - 15e^{2a}}{(12b^5 e^{2b+2a} - 12a^5 e^{2b+2a})} \quad (16)$$

The dissociation energy D of the bond can be computed by subtracting the zero point energy $E(\psi_{11})$ from the depth of the well.

The force constant of the bond can be found by taking the second derivative of the potential energy function. This potential, named after physicist Philip M. Morse, is a convenient model for the potential energy of a diatomic molecule and it's often transferred as a component of more complex structures. It is a better approximation for the Vibrational structure of the molecule than the quantum harmonic oscillator because it explicitly includes the effects of bond breaking, such as the existence of unbound states. Moreover, it accounts for the anharmonicity of real bonds and the non-zero transition probability for overtone and combination bands. However, this potential gives very small restoring forces for large ψ_{12} and, consequently, it causes slow convergence when employed in the optimization of the molecular geometry. For this reason, the truncated polynomial expansion is usually preferred.

With the Morse potential, the Eigen values of the stationary states are given

$$E(\Lambda) = hv_0\left(\Lambda + \frac{1}{2}\right) - \frac{(hv_0)^2\left((\Lambda + \frac{1}{2})\right)^2}{4D} \tag{17}$$

where Λ is the Vibrational quantum number and v_0 is dimensionally an energy, that can be obtained by the particle mass m and the Morse constants D and ρ with the formula

$$v_0 = \sqrt{\frac{k}{m}} = \rho\sqrt{\frac{2D}{m}} \tag{18}$$

Bending Energy Bending energy potentials can be treated very similarly to stretching potentials; the Taylor series can be truncated with n = 2 and yields the formulation

$$E_b\left(\theta_{12} - \theta_0\right) = c_2(\theta_{12} - \theta_0)^2 \tag{19}$$

Figure 4.4 shows the bending action on a toy structure. Hence, the energy is assumed to increase quadratic ally with angular displacement of the bond angle θ_{12} from equilibrium θ_0(Fig. 4.5).

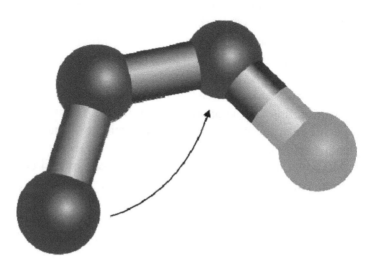

FIGURE 4.4   Representation of a bending action.

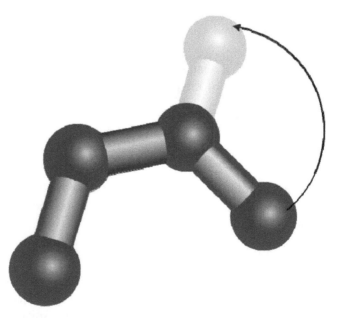

FIGURE 4.5   Representation of a Torsional action.

However, not only the energy must be periodic

$$E_b(\theta) = E_b(\theta + 2\pi),$$

(20)

But an additional constraint must be specified in more refined approaches: when $\cos(\theta_{12} - \theta_0) = 1$, the derivative of the potential needs to go to zero. This condition can be enforced with a more specific potential formulation. Even if bending is involved, the potential for moving an atom out of a plane is sometimes treated separately from bending. An out-of-plane coordinate ξ is considered in out-of-plane bending and the specific potential is usually taken in the harmonic approximation and added to E_b. Again, its formulation is

$$E_0(\xi) = q2\xi^2$$

(21)

Torsional Energy applies to capture some of the electrostatic and static non-bonded interactions; the Torsional energy term is added to the general formulation. It attempts to describe the interactions between two atoms a_1 and a_4, which are connected through an intermediate bond between the atoms a_2 and a_4, in a structure that presents the form $a_1-a_2-a_3-a_4$, as visualized by Fig. 4.6. More formally, let ω be the angle between two planes, the first identified by the positions of the atoms a_1, a_2, a_3 and the second by the positions of the atoms a_1, a_2, a_3.

FIGURE 4.6   Representation of ethene (left) and ethane (right).

The Torsional potential is not expanded as a Taylor series because the Torsional angle ω can go far from equilibrium. Instead, Fourier series are used and lead to the following approximate representation:

$$E_t(w) = U_p + \sum_{i=1}^{n} V_n \cos(j\omega).$$

(22)

\

The most common truncations are obtained with $n= 3$ or $n= 4$ (for octahedral complexes in organic chemistry) and the energy must be non-negative (imposing this condition with an appropriate Up). In the former truncated representation, one has the following form:

$$E_t(w) = \frac{1}{2}V_1\left(1+\cos(\omega)\right) + \frac{1}{2}V_2\left(1-\cos(2\omega)\right) + \frac{1}{2}V_3\left(1+\cos(3\omega)\right),$$

(23)

whose components are separately plotted in Fig. 4.7. For molecules like ethylene (official IUPAC name ethene), represented on the left-hand side of Fig. 4.6, the central C = C bond must be periodic. As the period is π, only even terms will exhibitnon-zero coefficients. Vice versa, for molecules like ethane, represented on the right-hand side of Fig. 4.6, only odd terms will presentnon-zero coefficients.

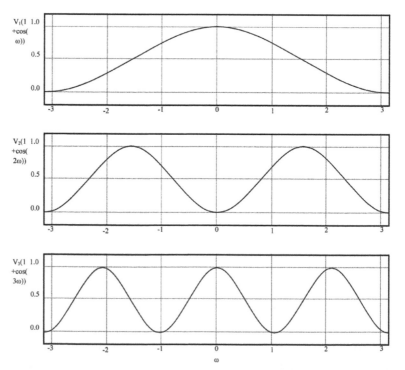

FIGURE 4.7   Separate plots of the three components of the energy.

Van der Waals Energy Interactions between the electron clouds around a couple of non-bonded atoms are the source of the van der Waals energy. When the range is short enough, this interaction is strongly repulsive. Vice versa, at the intermediate range, the interaction is attractive. As the distance of the atoms $\psi_{14} \rightarrow +\infty$, the interaction of course vanishes. The attraction is due to the electron correlation: a fluctuation of the electrons on one atom produces a temporary dipole,which induces a complementary dipole on the other atom. The resulting attractive force is called a dispersion (or "London") force van der Waals energies are usually computed for atoms which are connected by no less than two atoms, as shown in Fig. 4.8. Interactions between atoms closer than this are already accounted for by stretching and bending terms. At intermediate to long ranges, the attraction is proportional to $1/\psi^6{}_{14}$. At short ranges, the repulsion is close to exponential. Therefore, an appropriate model of the Van der Waals interaction is given by

$$E_v\left(\psi_{14}\right) = Ce^{-D\psi_{14}} - \frac{U}{\psi_{14}^6} \tag{24}$$

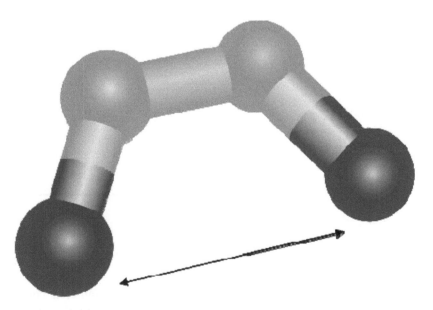

FIGURE 4.8   Representation of a non-bonded interaction.

However, a technical problem with the above"Buckingham" or "Hill" potential is that

$$\lim_{\psi_{14} \to 0} E_v\left(\psi_{14}\right) = -\infty \tag{25}$$

Since the van der Walls interaction is a long-range physical phenomenon, it becomes the dominant cost of a FF computation, which is usually quadratic on the number of atoms. This is the reason why a more economical approximation can substantially speed up the computation; this approximation is the Lennard-Jones potential [230], plotted in Fig. 4.9 and defined as:

$$E_{LJ}\left(\psi_{14}\right) = 4\varepsilon\left(\left(\frac{\psi_0}{\psi_{14}}\right)^{12} - \left(\frac{\psi_0}{\psi_{14}}\right)^6\right), \tag{26}$$

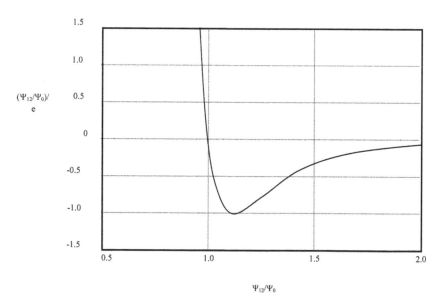

FIGURE 4.9   Plot of the normalized Lennard-Jones potential energy.

where ε is the dislocation energy and ψ_0 is the collision diameter. The first term, which involves $1/\psi^{12}_{14}$, is convenient for the computation because it's a simple polynomial approximation. The form of the repulsion term has a numerical but not a chemical justification: the repulsion force should depend exponentially on the distance, but the repulsion term of the formula is computationally faster due to the ease and efficiency of computing $1/\psi^{12}_{14}$ as the square of $1/\psi^6_{14}$.

The second term is the attractive long-range potential and it derives from dispersion interactions. Alternatively, it's possible to use a Morse potential with much smaller values for D, for ρ and a larger value for ψ_0. Non-bonded interactions between hydrogen and nitrogen or oxygen are definitely stronger (1 to 5 kcal/mol) than normal van der Waals interactions (0.1 to 0.2 kcal/mol) [231] and can be treated as special hydrogen stretching terms. Often, the extraction of these parameters comes by fitting of experimental data or by deduction from results of more precise and computationally expensive quantum calculations in chemistry.

The Lennard-Jones potential is often considered a good approximation, due to its simple form and it's often used to simulate the properties of gases, and to model dispersion and overlap interactions in molecular models. Due to its form, it is particularly accurate for atoms in noble gases and tends to be a good compromise at long and short distances for neutral atoms and molecules.

However, small deviations from the accurate empirical potential can be seen and are mostly due to the incorrect long-range part of the repulsion term. It's interesting to note that the minimum energy configuration of an infinite number of atoms described by a Lennard-Jones potential is a hexagonal arrangement of close-packed atoms, when the absolute temperature of the system tends to zero. Vice versa, raising the temperature of the molecular system, the minimum free energy arrangement becomes a cubic arrangement of close-packed atoms and finally becomes liquid. Under pressure, the lowest energy structure alternates between cubic and hexagonal arrangements [231].

4.4 ELECTROSTATIC ENERGY

The Coulomb interaction between the atoms a_1 and a_4 with partial charges, as seen in Fig. 4.10, is described by the electrostatic term in the global energy calculation. The formulation is as follows

$$E_e\left(\psi_{14}\right) = \frac{Q_1 Q_2}{\varepsilon \psi_{14}}, \tag{27}$$

where Q is an empirical dielectric constant whose value is unitary in vacuum but higher when there are intermediate atoms or in the presence of a solvent. The value of Q is obtained from fitting of experimental data but, in some cases, higher values or "distance-dependent dielectrics" with the form

$$\grave{o}_{dd} = \epsilon_0 \, \psi_{14}$$

(28)

where ϵ_{dd} can substitute the basic formulation, in order to simplify the truncation of the negligible interactions and to make the computation faster. Electrostatic terms are important in carbonyls and in other similar structures, where the carbon atoms have a partial positive charge and oxygen are partially negative. Also, hydrogen bonding is also sometimes accounted for by partial charges, even if, from the quantum mechanical point of view, it is hard to rigorously and unambiguously define atomic charges.

4.5   CROSS TERMS

Cross terms are sometimes required to account for some effects of the interrelation of different kinds of interactions, like the coupling of bending and stretching. A common and significant instance comes from the case of a strongly bent H_2O molecule. The bending action brings the two H atoms closer but the strain is partly relieved by the two O–H bonds that stretch and become a fraction longer than normal. Cross terms can model these interactions with the formulation that follows:

$$E_c\left(\psi_{123}\right) = E_{s/b}\left(\psi_{12}, \psi_2, \theta_{123}\right) = k_{123}\left(\theta_{123} - \theta_0\right)\left[\left(\psi_{12} - \psi_{12,0}\right) + \left(\psi_{23} - \psi_{23.0}\right)\right], \quad (29)$$

where θ_{123} is the angle on the atom a_2 between the atoms a_1 and a_3, $\psi_{12,0}$ is the equilibrium distance of the bond between a_1 and a_2, and similarly for ψ_{23}. Cross terms can be formed by different combinations that can represent stretch, bend-bend, stretch-torsion, bend-torsion and other more sophisticated interactions.

4.6   SPECIFIC POTENTIALS

A number of alternative potentials exist; they model specific conditions that depend on the particle type and on the characteristic distance. The interaction between two carbon atoms on the same CNT can be modeled by an effective short range potential of Tersoff-Brenner form [232].

The short range potential, which describes the covalent bonding, was given by Brenner [233]. Its parameters were given by Tersoff [232] to explain diversified forms of carbon, including diamond and graphite. However, some of these parameters can be modified in order to get a better fit for the equilibrium bond lengths in CNTs, such that the curvature of the surface can be accounted for. The potential energy between the atoms a_i and a_j on the same tube separated by a distance r_{ij} has the form that follows

$$V_B\left(r_{ij}\right) = f_c\left(r_{ij}\right)\left[f_R\left(r_{ij}\right) - b_{ij}f_A\left(r_{ij}\right)\right], \tag{30}$$

where the auxiliary functions are

$$f_R\left(r\right) = Ae^{-\lambda_1 r}$$

$$f_A\left(r\right) = -Be^{-\lambda_2 r} \tag{31}$$

and λ_1, λ_2, A and B are real parameters. In the formulation Eq. (4.30), f_c is a cut-off function, which is often taken as

$$f_c\left(r\right) = \begin{cases} 1, r < \left(R - D\right) \\ \dfrac{1}{2} - \dfrac{1}{2}\sin\left(\dfrac{1}{2}\pi\dfrac{r - R}{D}\right)\left(R - D\right) \leq r \leq \left(R + D\right) \\ 0, r > \left(R + D\right) \end{cases} \tag{32}$$

and is represented in its normalized form in Fig. 4.10. This form of cutoff, which goes from 1 to 0 in a small range around R, is continuous and has a derivative for all $r \in R$. The real parameter $R \in R$ is chosen to include only the first-neighbor shell, that is a $CC < R < \sqrt{3}aCC$.

The term b_{ij} implicitly includes the bond order and must depend on the local atomic environment. All deviations from a simple pair potential are accounted by the dependence of b_{ij} on the local atomic environment.

Precisely, the bonding strength b_{ij} for the pair of atoms a_i and a_j is a monotonically decreasing function with the following form

$$b_{ij} = \frac{1}{\sqrt[2n]{1 + \beta^n \xi_{ij}^{n}}}$$ (33)

where

$$\xi_{ij} = \sum_{k \neq j} f_c\left(r_{ij}\right) g(\theta_{ijk}) e^{\lambda_3^3 (r_{ij} - r_{ik})^3}$$

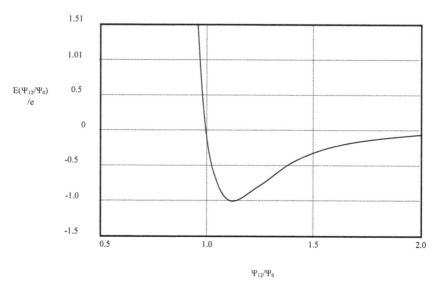

FIGURE 4.10   Plot of the cut-off function in the Brenner potential.

with k over all the neighboring atoms of a_i, θ_{ijk} is the angle between the bonds b_{ij} and b_{ik} and

$$g(\theta) = 1 + \frac{c^2}{d^2} - \frac{c^2}{(d^2 + (h - cos\theta)^2)}$$ (34)

The specific parameters [232] are the following

$$A = 1393.6 \ eV$$

$$B = 346.7 \ eV$$

$$\lambda_1 = 3.4879\text{Å}^{-1}$$

$$\lambda_2 = 2.2119\text{Å}^{-1}$$

$$R = 1.8 \ \text{Å}$$

$$D = 0.3 \ \text{Å} \ (435)$$

$$\beta = 1.5724. \ 10^{-7}$$

$$n = 0.72751$$

$$c = 3.8049. \ 10^{-4}$$

$$d = 4.3484$$

$$h = -0.577058$$

$$J_1 = \sqrt{3}i_1$$

Extraction of the Parameters from Experimental Data Modeling by MM requires a large number of parameters. Moreover, the quantity of potentially useful specific parameters may grow polynomial (but rarely linearly) with the number of atoms involved in the molecule. Often, only the most useful are considered and the literature of MM2 contains a large but limited set.

For instance, the number of torsion parameters has 2466 entries in MM2 but that means also that some (maybe exotic) torsions cannot be described properly without the introduction of the fitted constants. This lack of parameters is a serious drawback of all FF methods. When the data is not available, of course, generic parameters cannot be used, since they could lead to inaccurate results. Furthermore, the extraction of the necessary parameters from experimental data is often challenging, especially because experiments usually probe molecules at their equilibrium geometries and in conditions that are not easy to identify. That's the reason why Ab initio electronic structure methods are commonly used, when possible, to determine some of the parameters. Unfortunately, reliable values for van der Waals interactions are arduous to obtain, except for the most computationally intensive Ab initio computations. In MM, most of the necessary measurements come from one or more of the following techniques: electron diffraction, X-ray and neutron diffraction, microwave spectroscopy and high-resolution spectroscopy.

4.7   EXTENSIONS AND HYBRID METHODS

There are many recent and specialized FF methods. Some of these contain high order terms, for instance can account for quartic terms in stretching potentials, and several types of cross terms. These extensions have higher accuracy but are generally confined to molecules with small or medium size. Examples of these include Aligner's MM1 to MM4 [234], EFF and CFF. For very large molecules, like proteins, it is not possible to afford these advancements because of the prohibitive computational cost in terms of both space and time. The FF methods can be made more affordable by using only quadratic Taylor expansions and neglecting the entire cross terms. This leads to FF methods such as AMBER [235], CHARMM [236], GROMOS 6 [237] and others. For the sake of efficiency, these methods natively consider small set of atoms, like CH_2 units, as single particles. Hybrid methods, like the ONIOM method, are extensions of the basic concepts that treat some portions of the molecules with FF methods and some other parts by more accurate electronic structure methods. This approach is useful for systems where different accuracies are needed by different parts of the modeled molecule. They can be employed also when in parts of the molecule no FF parameters are known and experiments are not straightforward. Other popular names for these methods include Quantum Mechanics/Molecular Mechanics (QM/MM) methods.

4.8   HOMOGENIZATION FROM GRAPHENE MODELING

Homogenization is a mathematical technique used to derive macroscopic models for materials with micro or nano-structures. Its application is broad and includes the modeling of the heat flow, of the elastic deformations of composite materials or of the fluid flow in porous media. This mathematical theory provides a large-scale equation from fine scale equations. Considering a sequence of different problems, each one with more and more fluctuations on a smaller and smaller length scale, in the limit the fluctuations disappear. As a result, it's possible to obtain equations with homogeneous coefficients, which can be often easily solved on the large scale. It's important to point out that this remarkable mathematical procedure, sometimes counterintuitive to the physical in tuition, can provide accurate models. In its application to nano metric problems, deriving the limit can be very challenging and advanced competences are required. Averaging and homogenization both assume a more or less homogeneous structure for the examined material. In some cases, this assumption fails in the presence of irregular atom vacancies or, more generally, in the presence of structures with long-range correlations. Such structures will be missed by a homogenization procedure that uses only local information and that's the reason why homogenization is often combined with other techniques that come from the specific subject of application. The following subsections present a homogenization procedure for graphene sheets that can be extended to SWCNTs [238].

4.9   LATTICE CONFIGURATION AND VARIATION FORMULATION

The definition of a reference configuration of a graphene sheet is based on the repetition of a Y-shaped elementary cell. In a physical two-dimensional space P, let O be an arbitrary origin point and let i_1 and i_2 be two orthonormal vectors. The generic point M is identified by the vector OM. An alternative couple of vectors J_1 and J_2 are introduced, in order to describe a regular hexagonal lattice.

The couple of vectors are given by

$$J_2 = \sqrt{3}\left(\frac{1}{2}i_1 + \frac{\sqrt{3}}{2}i_2\right).$$

(36)

For a given $r \in R^{+\square}$, a regular hexagonal lattice with side length r can be described by introducing the scaled coordinate system given by

$$j_1 = rJ_1,$$

$$j_2 = rJ_2.$$

(37)

The grid consisting of points with integer coordinates along j_1, j_2 consists of half of the lattice vertices in a graphene sheet at rest. It can be described as the set of grid points

$$M\left(v_1, v_2, r\right) = v_1 j_1 + v_2 j_2, \left(v_1, v_2\right) \in Z^2 \tag{38}$$

An additional shift vector S must be introduced, in order to define the set of grid points that constitute the remaining half of the graphene sheet.

$$S = k\left(\frac{\sqrt{3}}{2} i_1 + \frac{1}{2} i_2\right)$$

Now, let assume that ε is small and that Θ^ε is a physical quantity defined on the ε-grid. The basic idea of a homogenization procedure is to assume that for $v\varepsilon$ kept fixed and equal to λ and ε going to 0, the expansion of Θ^ε follows

$$\theta^\varepsilon\left(M\left(v_1, v_2, \varepsilon r\right)\right) = \sum_{i=0}^{+\infty} \varepsilon^i \theta^i\left(\lambda_1, \lambda_2\right). \tag{39}$$

The definition leads to an elementary Y-shaped cell that consists of two atoms and of three bonds. Given a label $v = (v_1, v_2)$, in the $(0, 0)$-cell, the first node is located in O and the second in O $+\varepsilon k$. For generic v-cells, the first node is numbered as $(1, v)$ and the second as $(2, v)$. Translations of the elementary cell permit the timing of the entire ε-grid, with the identification of the nodes and their respective labels (n, v), with

$$(n, v) \dot{\mathrm{o}} \bar{N} = N_R \times Z^2 = \{1, 2\} \times Z^2. \tag{40}$$

The bonds $\tilde{b} = (b, v)$ can be numbered in a similar way, with

$$(b, v) \dot{\mathrm{o}} B_R \times Z^2 = \{1, 2, 3\} \times Z^2 \tag{41}$$

and each bond connects an origin node $\tilde{O}(\tilde{b})$ with an end node $\tilde{E}(\tilde{b})$, whose labels are respectively given by $(O(b), v)$ and $(E(b), v)$. The general forms for the definition of the bond relations are given by

$$\tilde{o}\left(\tilde{b}\right) = \left(1, v_1 - \delta_1\left(b\right), v_2 - \delta_2\left(b\right)\right),$$

$$\tilde{E}\left(\tilde{b}\right) = \left(2, v_1 - \delta_1\left(b\right), v_2 - \delta_2\left(b\right)\right). \tag{42}$$

Likewise, pair of bonds that share a common atom are numbered as $\tilde{c} =$ (c, v), where $c \in C_R = \{1, 2, 3, 4, 5, 6\}$. The first bond is denoted by

$$\tilde{P}(c) = \big(P(c), v_1, v_2\big) \tag{43}$$

and belongs to the same Y-cell as $\tilde{c}$, while the second

$$\tilde{D}(\tilde{c}) = \big(D(c)\big), v_1 + \delta_5(c) - \delta_6(c), v_2 - \delta_4(c) + \delta_6(c)\big), \tag{44}$$

where the auxiliary functions are defined as

$$P(c) = \begin{cases} 1, c \bar{o} \{3,6\} \\ 2, c \bar{o} \{1,4\} \\ 3, c \bar{o} \{2,5\} \end{cases} \tag{45}$$

and as

\

$$D(c) = \begin{cases} 1, c \bar{o} \{2,5\} \\ 2, c \bar{o} \{3,6\} \\ 3, c \bar{o} \{1,4\} \end{cases} \tag{46}$$

To each atom $(n, v) \in \tilde{\ } B$ of a ε-network, a reference position is assigned. It follows

$$\psi_0^\varepsilon(1,v) = \varepsilon\big(v_1 j_1 + v_2 j_2\big),$$

$$\psi_0^\varepsilon(2,v) = \varepsilon\big(v_1 j_1 + v_2 j_2\big) + \varepsilon r e_{30}, \tag{47}$$

where

$$e_{10} = -\frac{\sqrt{3}}{2} i_1 + \frac{1}{2} i_2,$$

$$e_{20} = -i_2, \tag{48}$$

$$e_{30} = -\frac{\sqrt{3}}{2} i_1 + \frac{1}{2} i_2$$

The deformed positions of the atoms $\tilde{O}\,(\tilde{b})$ and $\tilde{E}(\tilde{b})$ are $\psi\,(\tilde{O}\,(\tilde{b}))$ and $\psi\,(\tilde{E}(\tilde{b}))$, while the remaining notation follows

$$B_{\tilde{b}}\left(\psi\right)=\psi\left(\tilde{E}\left(\tilde{b}\right)\right)-\psi\left(\tilde{o}\left(\tilde{b}\right)\right),$$

$$l_{\tilde{b}}\left(\psi\right)=\left\|B_{\tilde{b}}\left(\psi\right)\right\|,$$

$$e_{\tilde{e}}\left(\psi\right)=\frac{B_{\tilde{b}}\left(\psi\right)}{l_{\tilde{b}}\left(\psi\right)},\tag{49}$$

$$P_{\tilde{c}}\left(\psi\right)=e_{\hat{P}_{(\tilde{c})}}\left(\psi\right).e_{\tilde{D}_{(\tilde{c})}}\left(\psi\right).$$

Recalling Section 4.1, a MM harmonic potential for bond stretching and angle bending is introduced. It is denoted by $W\colon \psi \to W(\psi)$ and its formulation is the following

$$W=\sum_{b\in B}\frac{k_{1}}{2}(l_{\tilde{b}}-r)^{2}+\sum_{\tilde{c}\in C}\frac{k_{p}}{2}(arccosp_{\tilde{c}}-\frac{2\pi}{3})^{2},\tag{50}$$

where k_{1} and k_{p} are the stiffness coefficients. Since the function arc $\cos(x)$ is not differentiable for $x\in \{-1, +1\}$, it's convenient to replace the entire

$$\left(\text{arc }\cos(x)-\frac{2\pi}{3}\right),$$

with another function h: $[-1, 1] \to R$ that approximates its behavior but it's differentiable for $x\in \{-1,+1\}$ [238]. Let h: $[-1, 1] \to R$ be defined as

$$\forall x \grave{o}[-1,1]\,h(x)\simeq\left(\text{arc }\cos(x)-\frac{2\pi}{3}\right)\tag{51}$$

then, for any v: $\tilde{N} \to R3$, the derivative of W is given by

$$W^{'}\left(\psi\right)(v)=$$

$$\sum_{\tilde{b}\in\tilde{B}}k_{i}\left(l_{\tilde{b}}-r\right)e_{\tilde{b}}.\left[v\left(\tilde{E}\left(\tilde{b}\right)\right)-v\left(\tilde{o}\left(\tilde{b}\right)\right)\right]-\sum_{\tilde{c}\in C}k_{p}\overline{h}\left(p_{\tilde{c}}\right)\left(e^{\frac{e_{\overline{D_{(c)}}}}{p(c)}}\right)e_{\tilde{b}}.\left(\omega_{\tilde{p}_{(\tilde{c})}}-\omega_{\overline{P}_{(\tilde{c})}}\right).\tag{52}$$

In the above formulation, the short notation imposes

$$l_{\bar{b}} = l_{\bar{b}}(\psi),$$

$$e_{\bar{b}} = e_{\bar{b}}(\psi),$$

$$p_{\bar{c}} = p_{\bar{c}}(\psi),$$ (53)

$$\omega_{\bar{b}} = \omega_{\bar{b}}(\psi, v)$$

$$= \frac{e_{\bar{b}}(\psi)}{l_{\bar{b}}(\psi)}\left(v\left(\tilde{E}(\tilde{b})\right) - v\left(\tilde{o}(\tilde{b})\right)\right),$$

for the sake of readability.

4.10   HOMOGENIZED LAW

Let $N^{\varepsilon}_{b}(\psi_b)$ and $M^{\varepsilon}_{c}(\psi_b)$ be defined on the ε-lattice as follows

$$N^{\varepsilon}_{\bar{b}}(\psi^{\varepsilon}) = k^{\varepsilon}_{l}\left(l_{\bar{b}}(\psi^{\varepsilon}) - r^{\varepsilon}\right)e_{\bar{b}}(\psi^{\varepsilon}),$$ (54)

$$M^{\varepsilon}_{\bar{c}}(\psi^{\varepsilon}) = -k^{\varepsilon}_{p}\bar{h}\left(p_{\bar{c}}(\psi^{\varepsilon})\right)e_{p(\bar{c})}\ e_{D_{(\bar{c})}}(\psi^{\varepsilon})$$

The vectors in Eq. (4.55) represent the tension and the moment vectors, respectively. Assuming that the deformations ψ^{ε} can be expanded in series in ε, one has

$$\forall \varepsilon \in R^{+*} \forall v \in Z^2 \mid v\varepsilon = \lambda$$ (55)

$$\psi^{\varepsilon}(n, v) = \psi^0(\lambda) + \varepsilon\psi^1_n(\lambda) + ...,$$

with $n = 1, 2$ and when $\varepsilon \to 0$. The elementary components of the constitutive

$$l_{\bar{b}}\left(\psi^\varepsilon\right) = \left\| B_{\bar{b}}\left(\psi^\varepsilon\right)_2 \right\|,$$

$$e_{\bar{b}}\left(\psi^\varepsilon\right) = \frac{B_{\bar{b}}\left(\psi^\varepsilon\right)}{l_{\bar{b}}\left(\psi^\varepsilon\right)},$$

$$p_{\bar{c}}\left(\psi^\varepsilon\right) = e_{\bar{P}_{(\bar{c})}}\left(\psi^\varepsilon\right).e_{\bar{D}_{(\bar{c})}}\left(\psi^\varepsilon\right)$$

(56)

and g(1, 0), g(2, 0) are two functions such that

$$g(1,.), g(2,.): R^2 \to R^3,$$

$$g(n,.): \lambda \to g(n, \lambda), n = 1, 2,$$

$$\forall \varepsilon \in R^{+*}, \forall v \in Z^2, g^\varepsilon(n, v) = \varepsilon^2 g(n, v\varepsilon).$$

(57)

The function g (λ) is defined as follows

$$g(\lambda) = \sum_{n \in N_R} g(n, \lambda)$$

(58)

While

$$S_\alpha^0 = \sum_{b \in B_R} N_b^0 e_b^0 \delta_\alpha(b) + \sum_{c \in C_R} M_c^0 \wedge \left(\frac{e_{D(c)}^0}{l_D(c)} \delta_\alpha(D(c)) - \frac{e_{P(c)}^0}{l_P(c)} \delta_\alpha(P(c)) \right)$$

(59)

The two introduced functions permit to write the strong form

$$\int_{R^2} S_\beta^0 \cdot \frac{\partial v}{\partial \lambda_\beta} d\lambda - \int_{R^2} g.v d\lambda = 0$$

(60)

The strong form can be written more compactly as

$$-div_\lambda S^0 = g,$$

(61)

that is the classical form of an equilibrium equation in CM. The mechanical interpretation permits the identification of the tensor S° with the first Piola-Kirchhoff stress tensor, under the reference configuration spanned by λ, which is the continuous variable of the model.

The equilibrium equation of the continuous medium must be complemented with the constitutive law for the non-linear model of the homogenized membrane. Hence, the following part summarizes the extraction of the membrane model. Finding the constitutive law means defining for all $\lambda \in R^2$ the values of the stress vectors $S^{\circ}_1(\lambda)$, $SO_2(\lambda)$, in R^3, associated with the derivative of ψ° at λ.

Finally one gets,

$$S_1^0(\lambda) = \hat{S}_1\left(\nabla_\lambda \psi^0(\lambda)\right)$$

$$= S_1\left(-\partial \lambda_1 \psi^0(\lambda) + Z(\lambda), -\partial \lambda_2 \psi^0(\lambda), Z(\lambda)\right),$$

$$S_2^0(\lambda) = \hat{S}_2\left(\nabla_\lambda \psi^0(\lambda)\right) \tag{62}$$

$$= S_2\left(-\partial \lambda_1 \psi^0(\lambda) + Z(\lambda), -\partial \lambda_2 \psi^0(\lambda), Z(\lambda)\right),$$

where

$$\nabla_\lambda \psi^0(\lambda) = \left(\partial \lambda_1 \psi^0(\lambda), \partial \lambda_2 \psi^0(\lambda)\right). \tag{63}$$

Here, $Z(\lambda)$ is provided in terms of $\nabla_\lambda \psi^0(\lambda)$ by the implicit equation

$$S\left(-\partial \lambda_1 \psi^0(\lambda) + Z(\lambda), -\partial \lambda_2 \psi^0(\lambda) + Z(\lambda), Z(\lambda)\right) = 0, \tag{64}$$

where

$$S = S_1 + S_2 + S_3 \tag{65}$$

Moreover, the Euler-Lagrange equation associated with the minimization of the continuous potential w is

$$S\left(-G_1 + Z, -G_2 + Z, Z\right) = 0. \tag{66}$$

4.11 CONTINUUM MODELING WITH THE EXPONENTIAL CAUCHY-BORN RULE

The central idea behind the continuous modeling by the use of the Exponential Cauchy-Born Rule (ECBR) is the extension of the methods of crystal elasticity to lattices of reduced dimensionality that is thin films in a three dimensional space or chains in a two-dimensional one. The standard Cauchy-Born Rule (CBR) is an approximation that comes from finite crystal elasticity and can be extended in the context of CM, when dealing with manifolds. This extension [239] provides the theoretical means to model the kinematics and the dynamics of curved lattices and its application is appropriate when dealing with CNTs [240]. The concept of the exponential map, which comes from differential geometry, generalizes the CBR, accounting for the fact that the lattice vectors are chords of a curved manifold. As a result, hyper elastic constitutive relations can be formulated from the lattice model for continua of reduced dimensionality.

Their obtained closed-form falls within the framework of CM, without the necessity of atomistic approaches. It's remarkable that, recently, the kinematical hypothesis expressed by the CBR has been proven to be a consequence of first principles of mechanics for a restricted class of two-dimensional lattices. This significant theoretical result promotes the status of the CBR from postulate to theorem [241].

4.11.1   THE CAUCHY-BORN RULE

The CBR is an approximation to the behavior of the atomic positions in a crystalline structure when this is subject to a small strain. Basically, it states that the positions of the atoms follow the overall strain of the material. It generally yields a reasonable approximation for Body-Centered Cubic (BCC) and Face-Centered Cubic (FCC) crystal groups but is not applicable to other or more complex structures, like diamond. Figure 4.11 shows the FCC and BCC crystal groups. The CBR formulates a finite deformation continuous model for a space-filling defect-free crystal structure, linking the atomistic deformation tothat of the continuous medium. In this framework, the lattice vectors are treatedas tangent vectors to the studied manifold and this approach will be generalizedby the ECBR. Several references present the details of the rule, some of whichwill be described also in this section. In order to define the continuum strainenergy density for a given deformation, a representative domain of solid-statematter, which has the same structure as a single crystal, is considered. Theobtained energy density is the energy of the portion of domain,which is subjectto the deformation divided by the volume of the involved region. In the followingsections, the following notational convections are used:

summation on repeatedindices is implied and contra variant indices act on forms, while covariant indicesact on vectors.

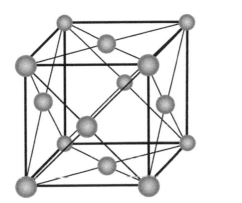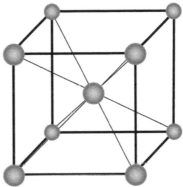

FIGURE 4.11   Exemplification of the FFC (left) and BCC (right) crystal groups.

Let Ω_0 be an under formed body in the n-dimensional Euclidean space, that is an open subset $\Omega_0 \subset R^n$, and let P be a point such that $X \in \Omega$. Then, the deformation φ maps Ω_0 into R^n and one has that

$$\Phi : X \to \Phi(X) = x.$$

The entire space-filling body 0 is deformed into $\Omega = \varphi(\Omega_0) \subset R^n$. According to the general definition [242], the deformation gradient F is the derivative of φ, which is a vector-valued vectorial function. It maps infinitesimal material vectors X to other material vectors x. It's possible to write that

$$F = D\Phi = \frac{\partial \Phi}{\partial P}$$

and, since it's a linear transformation from R^n to itself, it's possible to specify that $F \in R^{n*n}$. In the context of differential geometry, the deformation gradient F is also denoted as $T\varphi$ and called the tangent map of the deformation φ.

Then, the deformation φ maps Ω_0 into Ω and $T\varphi = F$ maps $T_x\Omega_0$ into $T_x\Omega$, where the former is the tangent space of 0 and the latter of. The standard CBR abstracts the method of the homogeneous deformations [243, 244], which is the central hypothesis behind molecular theories of elasticity at finite strains. More precisely, the latter predicts that, at the scale of the atomic spacing, the deformation of the crystal is homogeneous at finite strains. As a result, the crystal structure deforms and its lattice vectors undergo a linear transformation.

Let V be a vector in an undeformed lattice and let v be the same vector in the deformed crystal, then

$$v = FV. \tag{67}$$

In the more general case of multi lattices, in addition to the parallel superposition of the deformation of the single lattices, a number of additional kinematic variables accounts for the relative shifts of the separate lattices [245, 246].

Using relations from CM, the geometry of the deformed lattice vectors can be obtained from the continuum deformation by the use of the Green deformation

$$C = F^T F. \tag{68}$$

Using standard relations, both the relative angle θ_{12} and the length of the deformed lattice vectors v_1, v_2 can be easily obtained as follows:

$$\cos(\theta_{12}) = \frac{V_1 \cdot (CV_2)}{\|v_1\|_2 \|v_2\|_2} \tag{69}$$

And

$$\|v_i\|_2 = \sqrt{V_i \cdot (CV_i)} \tag{70}$$

for $i = 1, 2$.

For low-dimensional solids in R^3, it's convenient to present the formalization that follows [247–248]. Let an open set $\subset R^2$ be a referential body on the parametric space R^2. Let a function

$$\varphi_0 : \bar{\Omega} \to R^3 \tag{71}$$

be the differentiable and invertible parameterization function, such that

$$\varphi_0 : \xi \to \varphi_0(\xi) = X \in \Omega_0.$$

Likewise, let the deformed body Ω be another 2-differential manifold in R^3 and let be parameterized on the same referential body Ω by the de-formed configuration given by the function

$$\varphi : \bar{\Omega} \to R^3 \tag{72}$$

such that

$$\varphi : \Omega \to \varphi(\xi) = x \ \in \Omega.$$

Then, the previously introduced map

$$\Phi : \Omega_0 \to \Omega$$

can be defined as the composition of the two parameterization functions as follows:

$$\varphi_0 \ o \ \varphi^{-1} = \Phi \tag{73}$$

such that

$$\Phi : X \to \varphi\left(\varphi_0^{-1}(X)\right) = x. \tag{74}$$

Let B = $\{i_1, i_2, i_3\}$ be the standard basis of the three-dimensional Euclid-ean space and let it the space be described by its Euclidean coordinates $\{x^1, x^2, x^3\}$. Likewise, let B = $\{\xi^1, \xi^2\}$ be the standard basis and $\{\xi^1, \xi^2\}$ the Euclidean coordinates that describe the referential body Ω. If φ^i_0 and φ^i are the i-th Euclidean component of the respective configurations, then the respective convected basis vectors are given by

$$g_1 = \frac{\partial \varphi^1}{\partial \xi^1} i_1 + \frac{\partial \varphi^2}{\partial \xi^1} i_2 + \frac{\partial \varphi^3}{\partial \xi^1} i_3,$$

$$g_2 = \frac{\partial \varphi^1}{\partial \xi^2} i_1 + \frac{\partial \varphi^2}{\partial \xi^2} i_2 + \frac{\partial \varphi^3}{\partial \xi^2} i_3.$$

$$g_1 = \frac{\partial \varphi^1}{\partial \xi^1} i_1 + \frac{\partial \varphi^2}{\partial \xi^1} i_2 + \frac{\partial \varphi^3}{\partial \xi^1} i_3,$$

$$g_2 = \frac{\partial \varphi^1}{\partial \xi^2} i_1 + \frac{\partial \varphi^2}{\partial \xi^2} i_2 + \frac{\partial \varphi^3}{\partial \xi^2} i_3. \tag{75}$$

Consequently, for all $X \in \Omega_0$, the convected basis of $T_X\Omega_0$ is given by

$$C_0 = \{G_1, G_2\}.$$

Also, the cotangent space $T^*_X\Omega_0$ is the space of the linear mappings from the tangent space $T_X\Omega_0$ into R. The corresponding dual basis

$$C_0^* = \left\{G^1, G^2\right\}$$

of $T^*_X\Omega_0$ is defined by the relation

$$G^\alpha(G_\beta) = \delta_{(\alpha,\beta)}. \tag{76}$$

Likewise, for all $x \in \Omega$, the convected basis of $T_X\Omega$ is given by $C = \{g^1, g^2\}$. The cotangent space $T^*_X\Omega$ is the space of the linear mappings from the tangent space $T_X\Omega$ into R. The corresponding dual basis

$$C^* = \left\{g^1, g^2\right\}$$

of $T^*_X\Omega$ is defined by the relation

$$g^\alpha\left(g_\beta\right) = \delta_{\alpha,\beta} \tag{77}$$

Introducing the dual basis $B^* = \{\xi1, \xi2\}$, the tangent maps of the configurationscan be expressed as follows:

$$T\varphi 0 = G_1 \otimes \xi^1 + G_2 \otimes \xi^2 \tag{78}$$

And

$$T\varphi = g_1 \otimes \xi^1 + g_2 \otimes \xi^2. \tag{79}$$

Using the definitions from Eqs. (4.78) and (4.79) with the chain rule, the deformation gradient

$$T\phi = T\varphi^0 T\varphi_0^{-1} \tag{80}$$

of the tangent map $\phi = \varphi \circ \varphi^{-1}_0$ can be specified as

$$g_1 \otimes G_1 + g_2 \otimes G_2 = F : T\Omega_0 \to T\Omega \tag{81}$$

and maps

$$W^1 G_1 + W^2 G_2 = W \rightarrow FW = w = W^1 g_1 + W^2 g_2.$$

4.11.2 THE EXPONENTIAL RULE

A fundamental difficulty is faced when dealing with solids of low-dimensionality: the essential distinction between the manifold and the tangent space, the former being curved and the latter being flat, doesn't allow the straightforward application of the CBR without a necessary extension. In the case of space-filling solids described with Euclidean geometry, this view of the lattice vectors as chords of the surface and not as tangent vectors is hidden.

In Riemannian geometry, an exponential map is a map from a subset of a tangent space $T_p M$ of a Riemannian manifold M to M itself. In a neighborhood of each regular point p of the manifold M, it is a differ morphism that is a map between manifolds which is differentiable and has a differentiable inverse. It intrinsically maps the tangent of a manifold into the manifold itself. The exponential map can be simply described as follows [249]:

"The exponential map $\exp_p$ at a point p in M maps the tangent space $T_p M$ into M by sending a vector v in $T_p M$ to the point in M a distance $\|v\|$ along the geodesic from p in the direction v."

Its definition is possible because of the existence and uniqueness of geodesics at any point, given a direction in the tangent space. For surfaces the extraction of geodesics in a particular coordinate system is not trivial as in the case of curves and requires the solution of a system of non-linear Ordinary Differential Equations (ODEs). The unknown variables of these ODEs, named Geodesic Differential Equations (GDEs), are the parametric coordinates of the geodesic and their coefficients are the Christoffel symbols of the surface [248].

Let X and Y be two point in Ω_0, the first being the tail and the second the head of the chord A, which denotes an undeformed lattice vector. Assuming Y and X to be as close as necessary such that $\exp_X$ is invertible at Y. At this point, the application of the inverse exponential map

$$exp_x^{-1}(Y) = W \tag{82}$$

yields the vector $W \in T_x\Omega_0$ that can be transformed into $w \in T_x\Omega$ by the deformation gradient F. Once the vector w is obtained, the exponential map can finally permit the determination of the point $z \in \Omega$ by

$$z = exp_x(w), \tag{83}$$

where $x = \phi(X)$. The resulting vector is the chord whose tail is x and whose head is z.

Composing the three main transformations, a specific point map can be denoted as F_X and defined [249] as follows:

$$F_x = exp_{\phi(x)} oF(X) oexp_X^{-1} \tag{84}$$

and

$$F_X : \Omega_0 \to \Omega$$

$$F_X : Y \to F_X(Y) = exp_{\Phi(X)}\left(F_{exp_x^{-1}}(Y)\right) = z, \tag{85}$$

Since, in general, $z6 = y = \phi(Y)$.

4.11.3   APPLICATION TO CARBON NANOTUBES

Since graphene sheets can be seen as monolayer multi lattice crystalline films, a continuous solid model for SWCNTs can be described as a surface. This two-dimensional body without thickness is, in fact, a two-manifold in R^3. There, the atoms lie on the surface of the tube and the lattice vectors are chords of that surface. The previous sections evidenced that it's not straightforward to define a homogeneous deformation, as required by the standard CBR. Moreover, theory shows [248] that uniform metric and uniform curvatures aren't necessarily compatible. The ECBR, however, provides a theoretical and formal way to overcome the mentioned difficulties. Its application requires the integration of the GDEs, whose exact solution requires very complex non-local models that are numerically solved by computationally expensive schemes. In fact, the deformed lattice vectors don't depend only on the deformation of the surface at a particular point but also in the neighborhood of the involved point. To make the numerical solution computationally feasible in space and time, an approximation to the exponential map can be used, which renders the model possibly local. In this section, the approximation considers the kinematic assumption a = $F_X(A)$ as the general framework for a family of extensions to the CBR in the case of the thin-film continuous model of a SWCNT.

The association of the continuum stretch to the atomistic bond stretch and of the continuum curvature to changes in bond angles is intuitive. However, this is not the case, since the final interdependence is non-linear. The consistency of the theoretical approach does not guarantee that given a boundary value problem, the continuous model will provide a good approximation to the atomistic solution. Until recent advancements [241], a rigorous analysis has been lacking even for the case of space-filling crystals.

As SWCNTs can be described as rolled graphene sheets, the reference system is a graphene sheet at rest. Being at rest, the sheet is at its equilibrium ground and occupies a perfectly planar region, which is an open set $\Omega_0 \subset R^2$. The body is mapped into a smooth surface $\Omega \subset R^3$ by the deformation map

$$\Phi = \varphi o \varphi_0^{-1}, \tag{86}$$

as defined in the previous sections. Then, $\phi(X)$ denotes the vector from the origin in R^3 to the point $x = \phi(X) \in \Omega$. Following the definitions of the previous section, one has also the equivalence

$$\Phi(X) = \varphi\left(\varphi_0^{-1}(X)\right), \tag{87}$$

where the components of each vector in the standard basis B coincide with the components of the point mappings. The tangent space $T_x\Omega$ is a linear space for all the points $x \in \Omega$ and the convected basis $C = \{g_1, g_2\}$ of the tangent of the deformed body T comes from Eq. (75). In the undeformed configuration, the matrix representation of the tangent map in the Euclidean bases B-B$_0$ is denoted as

$$[T\varphi_0]_{\bar{B}B_0}$$

and its components can be computed as

$$(T\varphi_0)_\alpha^A = \frac{\partial \varphi_\alpha^A}{\partial \xi^\alpha} \tag{88}$$

The element of area in the undeformed body can be expressed as

$$d\grave{U}_0 = det\left[T\varphi_0\right]_{\bar{B}B_0} d\xi d\xi^2,$$ (89)

since both B and B_0 are Euclidean bases. The deformation gradient can be written as

$$F = T\Phi = T\varphi o \varphi_0^{-1}$$ (90)

and, also, one has that

$$\left[T\varphi\right]_{C\bar{B}} = I_2,$$ (91)

where I_2 is the 2 * 2 identity matrix. The components of the deformation gradient in the bases B0-C are given by

$$F_A^\alpha = \delta_{\alpha,1}(T\varphi^{-1})_A^1 + \delta_{\alpha,2}(T\varphi^{-1})_A^2 = (T\varphi_0^{-1})_A^\alpha$$ (92)

In the convected basis C, the metric tensor of the surface is represented in the following matrix form:

$$[g]c = \begin{bmatrix} g_{11} & g_{12} \\ g_{21} & g_{22} \end{bmatrix} = [g]_C^T,$$ (93)

where the covariant components are obtained from the vectors g_1 and g_2 as

$$g_{\alpha\beta} = g_\alpha \cdot g_\beta = (g_\alpha)^1 (g_\beta)^1 + (g_\alpha)^2 (g_\beta)^2.$$ (94)

In the convected coordinates, the first fundamental form

$$I(W) = \|W\|, W \in T\Omega$$

can be expressed as

$$I(w) = g_{11}\omega^1\omega^1 + g_{12}\omega^1\omega^2 + g_{21}\omega^2\omega^1 + g_{22}\omega^2\omega^2,$$ (95)

Since

$$w = \omega^1 g_1 + \omega^2 g_2.$$

The matrix representation of the Green deformation tensor, which is defined as the pull-back of the metric tensor $C = \phi^*g$, is

$$\left[C^b\right]_{B_0} = [F]^T_{CB0} [g]_C [F]_{CB0}$$

$$= \left[T\varphi_0\right]^{-T}_{BO\bar{B}} [g]_C \left[T\varphi_0\right]^{-T}_{BO\bar{B}} = \left[C^b\right]^T_{B_0} \tag{96}$$

The unit normal to the deformed body Ω is

$$n = \frac{1}{\|g_1 \times g_2\|} g_1 \times g_2. \tag{97}$$

The matrix elements of

$$[k]_c = \begin{bmatrix} k_{11} & k_{12} \\ k_{21} & g_{22} \end{bmatrix}_C$$

are the covariant components of the curvature tensor and can be expressed as

$$k_{\alpha\beta} = n^1 \left(\frac{\partial g_\alpha}{\partial \xi^\beta}\right)^1 + n^2 \left(\frac{\partial g_\alpha}{\partial \xi^\beta}\right)^2 + n^3 \left(\frac{\partial g_\alpha}{\partial \xi^\beta}\right)^3 \tag{98}$$

and permit the formulation of the second fundamental form

$$II(w) = k_{11}\omega^1\omega^1 + k_{12}\omega^1\omega^2 + k_{21}\omega^2\omega^1 + k_{22}\omega^2\omega^2$$

The matrix expression in B0 of the pull-back of the curvature tensor $K = \phi^*k$ is given by

$$[k]_{B_0} = \left[T\varphi_0\right]^{-T}_{B\bar{B}} [k]_C \left[T\varphi_0\right]^{-1}_{BO\bar{B}} \tag{99}$$

The Eigen value problem

$$[k]_C [v]_C = k [g]_C [v]_C. \tag{100}$$

permits the extraction of the principal curvatures and directions in convected coordinates. Alternative formulations and detailed solutions are given in Ref. [250]. In the application of the ECBR, the undeformed body is conveniently considered to be flat, that is

$$\Omega_0 \subset R^2$$

Recalling that A denotes an undeformed lattice vector and a the vector after the deformation according to the linear transformation of the standard CBR seen in Eq.(68), the ECBR can be expressed as

$$a = \exp o\ FA. \tag{101}$$

As said before, the evaluation of the exponential map requires the knowledge of the geodesic curves. These can be obtained by integration of a system of two non-linear ODEs, whose coefficients are the Christoffel symbols.

A general closed-form solution is not available but several numerical methods are known with their drawbacks [250]. These are based on the easy extraction of the closed-form solutions in the case of cylinders.

When applied to arbitrary deformations of crystalline thin films, the exponential map can be approximated by decoupling of the principal directions, instead of building a local representation based on the local deformation at each point in the membrane. Let V_1 and V_2 be the two principal directions of the curvature tensor and w the tangent deformed lattice vector. The two corrections for w can be obtained from the closed-forms for assumed cylinders of radii $1/k_1$ and $1/k_2$. After the computation of the corrections, these are added in each direction and permit the extraction of a. For the planar under formed crystal with $\Omega_0 = T\Omega_0$, the ECBR is described by

$$a = \exp_{\Phi(X)} o\ F(X)A. \tag{102}$$

The originally planar membrane is modified by a deformation map. Its local deformation can be characterized by the Green deformation tensor C and the curvature tensor K in Ω_0. Let $\{\tilde{x}_1, \tilde{x}_2, \tilde{x}_3\}$ be an auxiliary Euclidean coordinate system centered at $x = \phi(X)$, whose axes are parallel to v_1, v_2 and $v_1 * v_2$. Then, the associated orthonormal basis is

$$\tilde{B} = \{v_1, v_2, v_1 \times v_2\}$$

and it differs from B only by a rigid body transformation. Its restriction to $T_x\Omega$ is $\{\tilde{x}1, \tilde{x}2\}$ with the basis $\tilde{B}_{Tx\Omega}\{v1, v2\}$. Then, let β be the angle such that

$$\cos\beta = \frac{V_1^T CA}{\sqrt{A^T CA}},$$

$$\sin\beta = \frac{V_2^T CA}{\sqrt{A^T CA}} \tag{103}$$

The components in the basis $\tilde{B}_{Tx\Omega}$ of the tangent deformed lattice vector w are

$$\begin{bmatrix} \omega^1 \\ \omega^2 \end{bmatrix} = \begin{bmatrix} w^T & v_1 \\ w^T & v_2 \end{bmatrix} = \begin{bmatrix} V_1^T & CA \\ V_2^T & CA \end{bmatrix} \tag{104}$$

In the first principal direction, the assumed cylinder C_1 can be isometrically parameterized from $T_x\Omega$ into R^3 as

$$C_1 : f_1\left(\tilde{x}^1, \tilde{x}^2\right) = \left(\frac{1}{k_1} \sin k_1\overline{\tilde{x}^1}; x^2, \frac{1}{k_1}\left(1 - \cos k_1\overline{\tilde{x}^1}\right)\right). \tag{105}$$

The geodesic $c(s)$, with arc-length parameter s, of C_1 that passes through x and is tangent to w is given by

$$C(s) = \left(\frac{1}{k_1} \sin\left(k_1 s \cos\beta\right); s \sin\beta; \frac{1}{k_1}\left(1 - \cos(k_1 s \cos\beta)\right)\right). \tag{106}$$

The evaluation of Eq. (105) at x, when

$$s = \|w\| = \sqrt{A^T CA}, \tag{107}$$

Yields

$$\left[exp_{x,c1}w\right]_B = \begin{bmatrix} \dfrac{1}{k_1}\sin k_1 w^1 \\ \dfrac{1}{k_1}\left(1 - \cos k_1 w^1\right) \end{bmatrix}. \tag{108}$$

This makes possible the explicit formulation of the exponential correction in the first principal direction, that is

$$[\Delta w_1]_{\tilde{B}} = [exp_{x,c1} w]_{\tilde{B}} - w = \begin{bmatrix} \dfrac{1}{k_1} \sin k_1 w^1 - w^1 \\ \dfrac{1}{k_1}(1 - \cos k_1 w^1) \end{bmatrix} \tag{109}$$

Similarly, for the second principal direction, the cylinder C_2 is parameterized as Follows

$$C_2 : f_2\left(\tilde{x}^1, \tilde{x}^2\right) = \left(\tilde{x}^1; \frac{1}{k_2}\sin k_2 \widetilde{x^2};, \frac{1}{k_2}\left(1 - \cos k_2 \widetilde{x^2}\right)\right), \tag{110}$$

while the respective exponential correction is given by

$$[\Delta w_2]_{\tilde{B}} = [exp_{x,c2} w]_{B} - w = \begin{bmatrix} \dfrac{1}{k_2} \sin k_2 \overset{0}{w^2} - w^2 \\ \dfrac{1}{k_2}(1 - \cos k_2 w^2) \end{bmatrix} \tag{111}$$

Finally, the ECBR is approximated by the map

$$a = FA + \Delta w_1 + \Delta w_2.$$

In the orthonormal basis ~B, the expression for the deformed lattice vector is

$$[a]_{\tilde{B}} = \begin{bmatrix} w^1 \dfrac{\sin k_1 w^1}{k_1 w^1} \\ w^2 \dfrac{\sin k_2 w^2}{k_2 w^2} \\ \dfrac{k_1 (w^1 \sin (k_1 w^1 / 2))^2}{(k_1 w^1)^2 / 2} + \dfrac{k_2 (w^2 \sin (k_2 w^2 / 2))^2}{(k_2 w^2)^2 / 2} \end{bmatrix} \tag{112}$$

This formulation leads to the calculation of the length a of the deformed bond a as

$$a = \sqrt{[a]_B^T [a]_B} \tag{113}$$

and to the calculation of the angle between a couple of bonds a and b as

$$\theta = arccos\frac{a^1b^1 + a^2b^2 + a^3b^3}{ab} \quad (114)$$

Consequently, the bond length and angles have been expressed in terms of continuum strain measures C and K. The obtained strain measures are expressed as

$$a = f(C, K; A),$$

$$\theta = g(C, K; A, B) \quad (115)$$

The undeformed body Ω_0 represents the graphene sheet; it has three different bonds A_{01}, A_{02} and A_{03}. These bonds can be represented as

$$[A_{01}]_{B0} = A_0 \begin{bmatrix} cos\theta_0 \\ sin\theta_0 \end{bmatrix},$$

$$[A_{02}]_{B0} = A_0 \begin{bmatrix} cos\left(\theta_0 + \dfrac{2\pi}{3}\right) \\ sin\left(\theta_0 + \dfrac{2\pi}{3}\right) \end{bmatrix},$$

$$[A_{03}]_{B0} = A_0 \begin{bmatrix} cos\left(\theta_0 + \dfrac{2\pi}{3}\right) \\ sin\left(\theta_0 + \dfrac{2\pi}{3}\right) \end{bmatrix}, \quad (116)$$

Where

$$|\theta_i| = \theta_0 \in \left(-\frac{\pi}{6}, \frac{\pi}{6}\right]$$

is the orientation and $A_0 = a_{CC}$ is the equilibrium bond length. This leads to the definition of the undeformed body for a nano tube of length L as

$$\Omega_0 = (0, L) \times (0, \pi d_t). \quad (117)$$

The initial deformed configuration of the SWCNT is defined in $\{x_1, x_2, x_3\}$ as

$$\Phi^1 = X^1,$$

$$\Phi^2 = \frac{d_t}{2} \cos \frac{2X^2}{d_t},$$

$$\Phi^3 = \frac{d_t}{2} \cos \frac{2X^2}{d_t}.$$

(118)

Particular attention must be paid due to the fact that graphene sheets are multilattices, consisting of two interpenetrating rhomboidal grids. Inner displacements are the relative shifts that are described by additional kinematic variables [246, 251, 252]. Let s be the shifting vector that denotes the inner displacements. In T_0, the undeformed lattice vectors become

$$A_1 = A_{01} + s,$$

$$A_1 = A_{01} + s,$$

$$A_1 = A_{01} + s,$$

(119)

due to the inner displacements. As a result, a given continuum deformation transforms the triplet of undeformed bond vectors according to the ECBR:

$$a_i = F_X(A_i) = F_X(A_{0i} + s), i = 1, 2, 3.$$

(120)

Recalling Eq.(114), the length of the deformed bond vectors can be formulated

As

$$a_i = f(C, K; A_i) = \overline{f}(C, K, s; A_{0i}), i = 1, 2, 3.$$

(121)

Likewise, the angle of a couple of deformed bonds that share a common atom can be formulated as

$$\theta_i = g(C, K; A_j, A_k) = \overline{g}(C, K, s; A_{0j}, A_{0k}), i = 1, 2, 3$$

(122)

and where $\{i, j, k\}$ is an even permutation of $\{1, 2, 3\}$. Considering a hexagonal representative cell of the graphene lattice, two nuclei are contained, one of each rhomboidal grid. The surface area of the cell is

$$S_0 = \left(\frac{3\sqrt{3}}{2}\right) A_0^2 \tag{123}$$

And the energy for unit area is the strain energy density. For the entire continuum membrane, its strain energy density can be obtained by the division of the energy of the cell by its area S_0. Recalling the Brenner potential [201, 253] seen in Eq.(30) and its relative formalism that expresses the energy in term of bond lengths and angles as a sum over the bonds, one has

$$E = \sum_i \sum_{j>i} \left[f_R(r_{ij}) - B_{ij} f_A(r_{ij}) \right]. \tag{124}$$

Alternatively, recalling the many-body expansion formalism of the MM2/MM3 models from Section 4.1, one has

$$E = \sum_{bonds} E_s(r_{ij}) + \sum_{angles} E_b(\theta_{ijk}, r_{ij}, r_{ik}). \tag{125}$$

Considering a representative cell, which is hexagonal for the graphene honeycomb lattice, two atoms are included and the strain energy density can be written, for the Brenner potential, as follows

$$W = W(C, K, s) = \frac{1}{S_0} \sum_{i=1}^3 \left[f_R(a_i) - B_{j,k} f_A(a_i) \right], \tag{126}$$

where S_0 comes from Eq. (122) and $\{i, j, k\}$ is an even permutation of $\{1, 2, 3\}$. The resulting hyper elastic potential depends on the stretch C, on the curvature K of the surface and on the displacement field s. At the constitutive level, the inner displacements can be eliminated. Under a given deformation of the lattice, the strain energy density can be minimized with respect to s as follows:

$$\hat{s}(C, K) = \arg(\min_s W(C, K, s)) \tag{127}$$

that implies

$$\frac{\partial W}{\partial s} \Big|_{s=\hat{s}} = 0. \tag{128}$$

The evaluation of the hyper elastic potential W(C, K, s)exists in closed-form, as seen in Eq.(125). Vice versa, the evaluation of

$$\widehat{W}(C,K)$$ (129)

requires the solution of a bivariate minimization problem, which can be numerically obtained by Newton's method, as described with details in Ref. [239].

The derivative of Eq. (128) with respect to the stretch can be computed in closed-form and it is given by

$$\frac{\partial \widehat{W}}{\partial C} = \frac{\partial W}{\partial C}\big|_{s=\hat{s}} = \left(\frac{\partial W}{\partial C} + \frac{\partial W}{\partial \partial s^A}\frac{\partial \widehat{s^A}}{\partial C}\right)\big| s = \hat{s}_0$$ (130)

The second Piola-Kirchhoff stress tensor can be defined as

$$S = 2\frac{\partial \widehat{W}}{\partial C} = 2\frac{\partial W}{\partial C}\big|_{s=\hat{s}}$$ (131)

Likewise, the Lagrangian bending tensor can be expressed as

$$m = \frac{\partial \widehat{W}}{\partial K} = \frac{\partial W}{\partial K}\big|_{s=\hat{s}}$$ (132)

Given KBC and CAB, the constitutive model permits the calculation of the strain energy density and the stresses. The principal curvatures are obtained as the solution of an alternative to the formulation presented in Eq.(99), more precisely

$$[K]_{B_0}[V]_{B_0} = k[C]_{B_0}[V]_{B_0}$$ (133)

i.e.,

$$K_{AB}V^B = kC_{AC}V^C,$$ (134)

and makes possible to find the principal curvatures and the pull-backs of the principal directions expressed in B_0. The inner relaxation is computed as the minimization of W(C, K, s) with respect to s and permits the extraction of the relaxed inner displacements ŝ, the relaxed strain energy density $\widehat{W}$ and the updated undeformed lattice as

$$\left(A_1\right)^A = \left(A_{01}\right)^A + \hat{s}^A,$$

$$\left(A_2\right)^A = \left(A_{02}\right)^A + \hat{s}^A,$$

$$\left(A_3\right)^A = \left(A_{03}\right)^A + \hat{s}^A.$$

(135)

Bond lengths, angles and their derivatives with respect to the strain measures are obtained by the application of the ECBR. Finally, the stress tensors are formalized as

$$S^{AB} = 2\sum_{i=1}^{3}\left(\frac{\partial W}{\partial a_i}\frac{\partial a_i}{\partial C_{AB}} + \frac{\partial W}{\partial \theta_i}\frac{\partial \theta_i}{\partial C_{AB}}\right),$$

$$m^{AB} = 2\sum_{i=1}^{3}\left(\frac{\partial W}{\partial a_i}\frac{\partial a_i}{\partial K_{AB}} + \frac{\partial W}{\partial \theta_i}\frac{\partial \theta_i}{\partial K_{AB}}\right).$$

The application of external forces on the atoms are accounted for by their continuous counterpart as body forces, whose corresponding total external energy is given by

$$\Pi_{ext}\int_{\Omega_0} B.\Phi d\Omega_0,$$

(136)

where B is the body force per unit undeformed area. If the external force is constant and applied on each atom, then B is given by

$$B = \frac{n}{S_0}f$$

(137)

Let ψ be a given deformation map and Ω_0 be a planar body, then the internal energy of an elastic membrane is given by

$$\Pi_{int}(\psi) = \int_{\Omega_0} \widehat{W}\left(C(\psi), K(\psi)\right)d\Omega_0.$$

(138)

The total potential energy of the system is given by

$$\Pi(\psi) = \Pi_{int}(\emptyset) - \Pi_{ext}(\psi) + \Pi_{nb}(\psi),$$

(139)

where $\Pi_{int}(\psi)$ is given by Eq. (138), $\Pi_{ext}(\psi)$ by Eq. (136) and $\Pi_{nb}(\psi)$ is the continuous counterpart of the total non-bonded energy. The latter takes the following form

$$\Pi_{nb} = \frac{1}{2}\int_{\Omega_0}\int_{\Omega_0 - Bx} V_{nb}\left(\left\|\Phi(X) - \Phi(Y)\right\|\right)d\Omega_{0Y}\,d\Omega_{0X}, \tag{140}$$

where B_X is a ball centered at X with a radius that is a function of the potential cut-off radius and Vnb is the continuous van der Waals energy double density. The non-bonded energy of the atomistic system, can be written as follows

$$E_{nb} = \sum_i \sum_{i<\in Bx} V_{nb}\left(r_{ij}\right), \tag{141}$$

where V_{nb} is the non-bonded potential, r_{ij} the distance between the atoms a_i and a_j, and B_i the set of atoms that are bonded to the atom a_i. The continuous van der Waals energy double density can be defined as

$$V_{nb}\left(d\right) = \left(\frac{n}{S_0}\right)^2 V_{nb}\left(d\right), \tag{142}$$

When two representatives cell each of area S_0 are presents. The stable equilibrium deformation maps are the minimizes of Π and, hence, they are given by

$$\Phi = ar\ g\left[\begin{matrix} inf\ \Pi(\psi) \\ \psi\partial C \end{matrix}\right] \tag{143}$$

where C is a suitable space of deformation maps accounting essential boundary conditions. The equilibrium configurations of the system ϕ are stationary points of the potential energy functional, according to the principle of stationary energy. These verify the principle of virtual work

$$\forall \delta\,\Phi \in V\delta\tilde{O}(\Phi) = \int_{\Omega 0}\left(\frac{1}{2}S:\delta C + m:\delta K\right)d\Omega_0 - \delta\Pi_{ext} + \delta\Pi_{nb} = 0, \tag{144}$$

where ϕ: stands for the double contraction of tensors. The variations of the non-bonded and external energy functional are given by

$$\delta \Pi_{nb} = \frac{1}{2} \int_{\Omega_0} \int_{\Omega_0 - Bx} \frac{1}{\Phi(X) - \Phi(Y)} V'_{nb} \left(\Phi(X) - \Phi(Y) \right) \left(\Phi(X) - \Phi(Y) \right).$$
$$\left(\delta\Phi(X) - \delta\Phi(Y) \right) d\Omega_{0Y} d\Omega_{0X} \tag{145}$$

and

$$\delta \Pi_{ext} = \int_{\Omega_0} B . \delta \Phi d\Omega_0. \tag{146}$$

KEYWORDS

- **Carbon Nanotubes**
- **Electrostatic Energy**
- **Homogenized Law**
- **Hybrid Methods**
- **Molecular Mechanics**
- **Variation Formulation**

CHAPTER 5

NUMERICAL SIMULATION OF THE MECHANICAL BEHAVIOR

CONTENTS

5.1   NUMERICAL SIMULATION OF THE MECHANICAL BEHAVIOR

5.1.1   PARAMETRIC MOLECULAR GENERATION

The high symmetry of SWCNTs permits the parametric generation of the atomic coordinates for any (n_1, n_2) tubule. The helical and rotational symmetries [254, 255] are used to define the full Euclidean symmetry group of infinite SWCNTs and, hence, to obtain the coordinates of the atoms. Line groups [255] are the group of Euclidean symmetries of the systems that exhibit translational periodicity in one direction. Typical examples of these structures are the quasi-one-dimensional crystals like SWCNTs. The monomers are the elementary structural units of the lattice and their regular arrangement is obtained by pure translations combined with operations on the screw axis. Monomers are clustered into larger local units: the elementary cells. Let a SWCNT be axially aligned on the z-axis in the Euclidean space and let P be an axial point group. Each line group L is a weak-direct product $L = ZP$ of a group Z of the generalized translations and the axial point group P. The former arranges the monomers, while the latter implies the symmetry of the monomers. The axial point group P leaves the z-axis invariant and the infinite cyclic group Z is either a screw axis or a glide plane group. In the Koster-Seitz notation, the generator of the glide plane group is denoted as

$$\left(\sigma_v \Big| \frac{a}{2}\right),\tag{1}$$

where a denotes the translational period of the group L and σ_v is the vertical mirror plane. The generator of the screw axis group, the latter denoted by $T^r_q(a)$, is

$$z = \left(C^r_q \Big| \frac{n}{q} a\right),\tag{2}$$

where q and r are non-negative integers such that $q = \alpha n$, for $\alpha \in N$. The choice of r is not unique, given r any multiple of q/n may be added, with no effect on the resulting group L. In order to establish a fixed value of r, two different conventions can be used [256]:

✓   r is coprime with q/n,
✓   r is the minimal allowed value that is coprime with q.

The translational period of L contains q/n monomers and each mono-
mer is obtained from the previous one by the rotation for the angle of
$2\pi q/r$, followed by the fractional translation of (q/n) a. infinitely many
line groups are possible and are classified into 13 classes, according to
the factors Z and P. In order to determine the line group that comprises all
the Euclidean symmetries of the nanotubes, the procedure of folding up a
graphene sheet at rest is used. The symmetries of this honeycomb lattice H
form the di periodic group $D_{g28} = D_{6h}T$. The translational group T is generated by the translations for the basis vectors a_1 and a_2. The principal axis of
order six of the group D_{6h}, which is perpendicular to H, passes through the
origin at the center of the hexagon. The elements of D_{g28} form the tube's
line group and are the symmetries of the rolled up lattice form. First, the
translations are described. For a (n_1, n_2) SWCNT, H is rolled up so that C_h
becomes the circumference of the tube. The translations of H along the
chiral vector become the rotations around the tube axis. The minimal allowed among these is given by

$$\hat{c} = \frac{C_h}{n}, \tag{3}$$

wheren = gcd $[n_1, n_2]$. Hence, the group of pure rotations of the tube is the
cyclic group C_n, generated by the rotation for $2\pi/n$. The pure translations
of the tube are the honeycomb translations in the directions orthogonal
to C_h, as previously stated for T_h. When n and Th are known, the screw
axis generator is found as follows. Each two-dimensional lattice translation becomes, on the surface of the tubule, an element of the group $T_q^r C_n$.
The honeycomb is generated by $T_q^r C_n$ from the pair of carbon atoms in the
honeycomb elementary cell, which contains q/n monomers. Each mono-
mer contains n elementary honeycomb cells, obtained by the action of C_n
and, also, there are q honeycomb cells in the tube's period. The area of the
latter cylindrical surface is $\|C_h\|_2 \, \|T_h\|_2$. Dividing the resulting area by the
area of the honeycomb elementary cell $\|a_1 \wedge a_2\|_2$, one finds

$$q = \frac{2\left(n_1^2 + n_1 n_2 + n_2^2\right)}{nR}, \tag{4}$$

where R = 3 if $(n_1 - n_2) = \alpha n$ for some $\alpha \in N$ and R = 1 otherwise. Therefore,
the rolled lattice is generated by the primitive translations a_1 and a_2 on the
surface that is the group $T_q^r C_n$ with elements given by

$$\left(C_q^{rti} C_n^s | t \frac{n}{q} a \right),$$
(5)

with $t \in Z$ and $s \in \{0, 1,..., n-1\}1$. The element that corresponds to a_i, for $i \in \{1, 2\}$ is given by

$$\left(C_q^{rti} C_n^s | t_i \frac{n}{q} a \right)$$
(6)

that corresponds to the rotation

$$\varphi_i = \frac{2\pi \left(rt_i + q^{\frac{si}{n}} \right)}{q}$$
(7)

followed by the translation for

$$T_i = \frac{t_i na}{q}$$
(8)

The parameters of these rotations are given by Ref. [140] and are

$$t_1 = -\frac{n_2}{n},$$

$$t_2 = -\frac{n_1}{n},$$

$$s_1 = \frac{2n_1 + (1 + rR) n_2}{qR},$$
(9)

$$s_2 = \frac{(1 - rR) n_1 + 2n_2}{qR}.$$

The minimal r that provides the solutions for s_1 and s_2 is co-prime to qn and, more precisely, it's given by

$$r = \frac{n_1 + 2n_2 - (\frac{n_2}{n})^{\varphi(n_2/n)-1} qR}{n_1 R} \mod \frac{q}{n},$$
(10)

where $\phi(m)$ is the Euler function, giving the number of co primes not greater than m.

Thus, the derivation of the symmetries of SWCNTs permits the parametric generation of the tubule by the described rotations and translations of the monomers, according to the rules that depend on the specified chirality. The implemented computer code permitted the generation of finite-length SWCNTs with any input chirality, as shown by the example in Fig. 5.1. The generated models are the molecular models used in the numerical simulations, which are presented in the following sections.

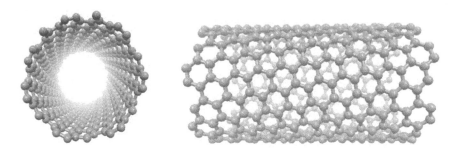

FIGURE 5.1   Generic chiral (13, 7) single-walled nano tube.

5.1.2 STRUCTURE OF THE MECHANICAL MODELS

5.1.2.1 REFERENCE CONFIGURATION

Reflecting the repetitive display of the lattice, a numbering scheme is presented, in order to define the discrete counterpart of the reference configuration in CM. The identification of the atoms with the vertices of the hexagons holds even when a mechanical deformation modifies the regular pattern. Therefore, the numbering can be performed with reference to the unstressed configuration of the sheet, obtaining a structure that describes the entire sheet as the tiling repetition of an elementary Y-shaped cell, as seen in Section 5.2. First, this modeling technique is applied to graphene [257] and, finally, to general SWCNTs. Formally, a sheet of graphene G is a set of carbon atoms, binary bonds and angles between adjacent bonds. It is represented by

$$C = (A, B, C) \tag{11}$$

where A is the set of all the atoms of the sheet, B is the set of all the binary bonds between pairs of adjacent atoms and C is the set of all the ordered couples of adjacent bonds. Every atom a ∈ A is defined by two attributes, so that it's possible to write

$$a = (pos_a, lab_a) \tag{12}$$

where

$$pos_a: [t_0 + \infty) \rightarrow R^3 \tag{13}$$

is a function of the time. These attributes are named, respectively, position at time t and label of the atom a. Let O ∈ R^3 be the position of a point that coincides with the center of a carbon atom of the graphene sheet in a physical three-dimensional Euclidean space. Also, let i_x, i_y, i_z be three orthonormal vectors, each with unitary length and with the additional constraint that the plane of the graphene sheet at rest coincides with span{i_x, i_y} = G. Hence, the position of every point P ∈ R^3 can be identified with the vector O→P and can be written as a linear combination of the vectors of the base {i_x, i_y, i_z}; then, their explicit linear combination is

$$P = \gamma_x i_x + \gamma_y i_y + \gamma_z i_z, \tag{14}$$

with the ordered triplet (γ_x, γ_y, γ_z) ∈ R^3. As a result, if at t = $\hat{t}$ the graphene sheet is at rest (that is, the molecule is at its ground energy) and

$$\forall a_i \in A pos_a\left(\hat{t}\right) \in G$$

It's possible to write the position of every atom a ∈ A with respect to the base {i_x, i_y, i_z} of (4.14) as

$$pos_a\left(\hat{t}\right) = [\gamma_x\left(\hat{t}\right), \gamma_y\left(\hat{t}\right), \gamma_z\left(\hat{t}\right)]^T \tag{15}$$

In Eq. (15), $\gamma_z(t) = 0$ as every point PG on the plane G of the graphene sheet is associated with the vector PG = [γ_x, γ_y, 0]^T, where γ_x, γ_y ∈ R. Because of the hexagonal structure of the lattice, another useful coordinate system that is not orthonormal can be specified for the physical space. Two new linearly independent vectors J_1, J_2 are introduced, such that

$$J_1 = \sqrt{3}i_x$$

$$J_2 = \sqrt{3}\left(\frac{1}{2}i_x + \frac{\sqrt{3}}{2}i_y\right).$$
(16)

Together with the previously introduced unitary vector i_z, a new base for R^3 is defined by $\{J_1, J_2, i_z\}$ and again one has that $G = \text{span}\{J_1, J_2\}$ holds. The Euclidean length of any of the two new vectors defined in Eq.(5.16) is

$$\|J_1\| = \|J_2\| = \sqrt{3}.$$
(17)

Their left angle amounts to

$$\prec\left(i_x, i_y, i_z\right)\left(J_1, J_2\right) = \frac{\pi}{3}.$$
(18)

As the length of the carbon-carbon bond $a_{(C-C)}$ is sub-nanometric, it's convenient to introduce the scaled versions of the previous vectors as follows:

$$j_1 = J_1 a\,(C - C)$$

$$j_2 = J_2 a\,(C - C)$$
(19)

$$j_3 = J_z a\,(C - C)$$

Whose lengths are $\sqrt{3}a_{(C-C)}$ for the first couple and $a_{(C-C)}$ for the third vector. Figure 5.2 shows the two grids and the presented vectors. A rhomboidal grid on the graphene sheet is defined as $A1 \subset A$:

$$A_1 = \left\{a \in A \mid pos_a\left(\hat{\imath}\right) = \alpha_1 j_1 + \alpha_2 j_2, \alpha_1, \alpha_2 \in Z\right\}$$
(20)

Every carbon atom $a \in A_1$ will be uniquely identified by the integer couple $(\alpha_1, \alpha_2) \in Z^2$ and will be referred to as

$$lab_a = A_1\left(\alpha_1, \alpha_2\right) = \left(\alpha_1, \alpha_2, 1\right).$$
(21)

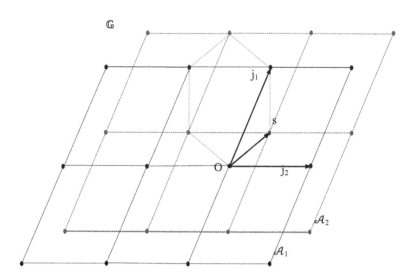

FIGURE 5.2 The two rhomboidal A_1 and A_2 showing also the shifting vector s and the vectors that compose the basis $\{j_1, j_2\}$ of G.

Let $A_2 = A/A_1$ be another rhomboidal grid with the same integer constraints. More precisely, A_2 is the set whose atoms have their positions that are the shifted version of the atoms in A_1. This shift is induced by the shifting vector s, defined as

$$s = a(C-C)\left(\frac{\sqrt{3}}{2}i_1 + \frac{1}{2}i_2\right) \tag{22}$$

that has length $a_{(C-C)}$. Hence,

$$A_2 = \left\{ a \partial A \mid pos_a\left(\hat{i}\right) = \alpha_1 j_1 + \alpha_2 j_2 + s, \alpha_1, \alpha_2 \in Z \right\} \tag{23}$$

The shifting vector s makes an angle with J_1 that amounts to

$$\prec\left(i_x, i_y, i_z\right)\left(J_1, s\right) = \frac{\pi}{6}.$$

As seen for the atoms in A_1 in Eq. (5.21), a similar numbering scheme holds for the atoms in A_2. The label for the atom $a \in A_2$ is given by

$$lab_a = A_2\left(\alpha_1,\alpha_2\right)=\left(\alpha_1,\alpha_2,2\right).\qquad(24)$$

The Y -shaped cell is the elementary tile that covers the entire graphene sheet, as visually represented in Fig. 5.3. Every cell is uniquely identified by the ordered pair of integers $(\alpha_1,\ \alpha_2)\in Z_2$ and consists of two atoms and three bonds. Hence, every cell Y $(\alpha_1,\ \alpha_2)$ is given by the quintuple

$$Y\left(\alpha_1,\alpha_2\right)=\left(a_1\left(\alpha_1,\alpha_2\right),a_2\left(\alpha_1,\alpha_2\right),b_1\left(\alpha_1,\alpha_2\right),b_2\left(\alpha_1,\alpha_2\right),b_3\left(\alpha_1,\alpha_2\right)\right),\qquad(25)$$

where the first two items are atoms and the remaining three are bonds. More precisely, $a_i(\alpha_1,\ \alpha_2)=(\alpha_1,\ \alpha_2,\ i)$ for $i=$ 1, 2 (Fig. 5.4).

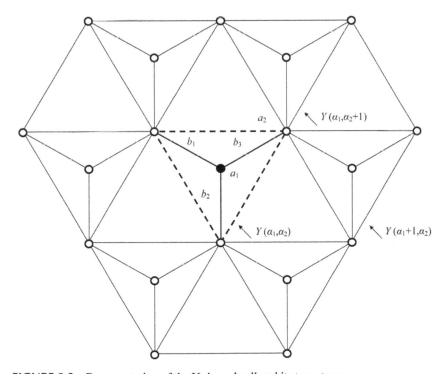

FIGURE 5.3 Representation of the Y shaped cell and its two atoms.

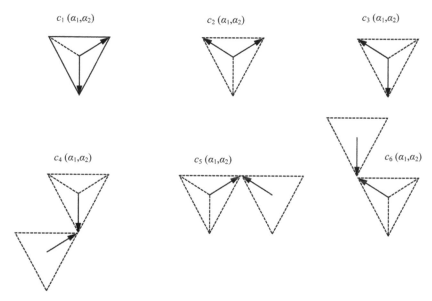

FIGURE 5.4   Representation of the numbering scheme for couples of bonds associated with the generic elementary Y-shaped cell Y (α_1, α_2).

Every bond can be specified by an ordered pair of neighboring atoms, so that, for every cell Y (α_1, α_2),

$$b_1\left(\alpha_1,\alpha_2\right)=\left(a_2\left(\alpha_1-1,\alpha_2\right),a_1\left(\alpha_1,\alpha_2\right)\right)$$

$$b_2\left(\alpha_1,\alpha_2\right)=\left(a_2\left(\alpha_1,\alpha_2-1\right),a_1\left(\alpha_1,\alpha_2\right)\right) \tag{26}$$

$$b_3\left(\alpha_1,\alpha_2\right)=\left(a_2\left(\alpha_1,\alpha_2\right),a_1\left(\alpha_1,\alpha_2\right)\right).$$

Vice versa, one can refer to the first atom of the bond $b_u(\alpha_1, \alpha_2)$ as b1 u (α_1, α_2) for $u = 1, 2, 3$ and similarly for the second atom.

In addition to atoms and bonds, the presented numbering scheme must be extended in order to include also couples of bonds that share a common atom. Six couples of bonds are associated to every Y-shaped elementary cell. Let a generic cell be Y (α_1, α_2) as usual; then these couples are named $c_i(\alpha_1, \alpha_2)$ with $i = 1, 2,...,6$ and can be described by the following ordered pairs of bonds:

$$c_1(\alpha_1,\alpha_2) = \left(b_2(\alpha_1,\alpha_2), b_3(\alpha_1,\alpha_2)\right)$$
$$c_2(\alpha_1,\alpha_2) = \left(b_3(\alpha_1,\alpha_2), b_1(\alpha_1,\alpha_2)\right)$$
$$c_3(\alpha_1,\alpha_2) = \left(b_1(\alpha_1,\alpha_2), b_2(\alpha_1,\alpha_2)\right)$$
$$c_4(\alpha_1,\alpha_2) = \left(b_2(\alpha_1,\alpha_2), b_3(\alpha_1,\alpha_2 - 1)\right) \qquad (27)$$
$$c_5(\alpha_1,\alpha_2) = \left(b_3(\alpha_1,\alpha_2), b_1(\alpha_1 - 1,\alpha_2)\right)$$
$$c_6(\alpha_1,\alpha_2) = \left(b_1(\alpha_1,\alpha_2), b_2(\alpha_1 + 1,\alpha_2 + 1)\right)$$

Vice versa, one can refer to the first bond of the couple $c_u(\alpha_1, \alpha_2)$ as $c_u^1(\alpha_1, \alpha_2)$ for $u = 1, 2,...,6$ and similarly for the second bond. The graphical representation of Eq. (27) is visualized in Fig. 5.5.

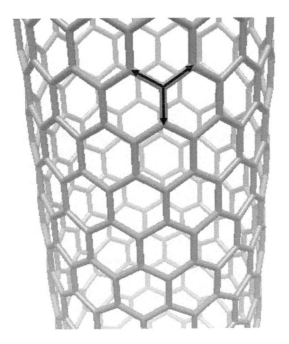

FIGURE 5.5 Applications of the tiling scheme on a SWCNT, highlighting a single Y-shaped element.

This numbering scheme successfully applies also to SWCNTs, where it is adopted according to the constraints deriving from the surface being cylindrical. Similarly to a graphene sheet, a SWCNT can be denoted by

$$S = (A, B, C). \tag{28}$$

Each chirality class of SWCNTs will show a different but regular tiling pattern of the structure. An example is shown in Fig. 5.5 and it's important to point out that the length of a(C–C) at rest has not the same value as in the graphene sheet; according to the pair (n_1, n_2), it varies with the direction on the plane that is tangent to the surface in $A1(\alpha_1, \alpha_2)$. As a result, SWCNTs at the mechanical equilibrium may show, in every Y-shaped elementary cell, different lengths of the three bonds. This issue will be discussed later with the minimization of the bond length at rest.

5.1.2.2   ENERGY AND DYNAMICS

Let L_s the set of atom labels in the SWCNT $S = (A, B, C)$. Then,

$$L_s = \left\{ lab_a \mid a \in A \right\} \tag{29}$$

Linearly ordering the atoms of S according to their labels, a triplet of vectors can describe the geometrical configuration of the system at a given time t. Let $N = |A|$; these three time-dependent vectors are

$$\forall t \in \left[t_0, + \infty \right) \forall a \in A \Phi_t \left(lab_a \right) = pos_a \left(t \right). \tag{30}$$

Linearly ordering the atoms of S according to their labels, a triplet of vectors can describe the geometrical configuration of the system at a given time t. Let $N = |A|$; these three time-dependent vectors are

$$\psi_x \left(t \right) = [X_1 \left(t \right), X_2 \left(t \right), \dots X_n \left(t \right)]^T,$$

$$\psi_y \left(t \right) = [y_1 \left(t \right), y_2 \left(t \right), \dots y_n \left(t \right)]^T, \tag{31}$$

$$\psi_z \left(t \right) = [z_1 \left(t \right), z_2 \left(t \right), \dots z_n \left(t \right)]^T,$$

and, for every triplet, the following equality holds

$$[X_i(t), y_i(t), z_i(t)]^T = pos_{ai}(t) \tag{32}$$

As seen before, several functional forms can be used for the energy terms, according to the different materials and loading conditions that are considered in the simulation. For SWCNTs, only the term Vi covalent is considered of interest when small strains are applied to the molecule, the other terms being negligible [258].

In such cases, only the terms accounting for bond stretching's and angle variations are significant in the system's potential. Hence, the resulting potential energy can be approximated under the assumption of small local deformations, the harmonic approximation for angles and bonds can be employed. This constraint is only local and doesn't forbid the simulation of large global deformations. Several calculations in chemistry have shown that a reasonable approximation to the potential energy of a molecular system can be provided by harmonic functions when the bond lengths are near to their equilibrium positions [259]. As it mentioned before, binary bonds are modeled as springs and their stretch is subject to Hooke's Law. The spring parameters are defined as solutions of best approximation problems of potential functions obtained from experimental data. They are the equilibrium distance $a_{(C-C)}$ and the spring constant k_l of linear elasticity. The same harmonic approximation scheme applies for angular springs: their equilibrium angle is $2\pi/3$ and their spring constant is k_p. When a SWCNT is subjected to external forces, the displacements of the individual atoms are constrained by bond stretching and angle bending. Hence, the global deformation is the combined result of these basic interactions.

In this numerical simulation, the values of the harmonic parameters are

$$k_1 = 652 \frac{N}{m}, \tag{33}$$

$$k_p = 8.76.10^{-19} \, Nm \tag{34}$$

These values are taken as granted by several authors [238] and have been introduced by [260].

As the equation of spring deformation can be used to describe the ability of bonds to stretch, in the introduced notation, let $b_t \in B_t$ be a bond of a SWCNT S. Then, by definition,

$$b = (b_1, b_2) \tag{35}$$

and its length is

$$l_b(t) = \|pos_{b2} - pos_{b1}\|_2 . \tag{36}$$

Therefore, the potential of a single bond stretch is given by

$$V_b(\psi_b(t)) = \frac{k_1}{2}(l_b(t) - a_{(C-C)})^2 . \tag{37}$$

Now, let $c \in C$ be a bond. Recalling its definition, one has that

$$c = (c^1, c^2) = ((b^1, b^2), (b^3, b^4)) \tag{38}$$

and the bonds must share a common atom. If the two bonds belong to the same Y-shaped cell, then the shared atom is $b_2 = b_4$; otherwise, $b_1 = b_3$. In both the cases, the angle $\rho_c(t)$ between the couple of bonds c is given by

$$\rho_c(t) = \frac{(pos_{b2}(t) - pos_{b1}(t))^T (pos_{b4}(t) - pos_{b3}(t))}{pos_{b2}(t) - pos_{b1}(t)_2 \, pos_{b4}(t) - pos_{b3}(t)_2} . \tag{39}$$

Adjacent bonds tend to maintain their equilibrium angles and, again, the angle bending potential for c is given by

$$V_c(\psi_c(t)) = \frac{k_p}{2}(acos\rho_c(t) - \frac{2\pi}{3})^2 \tag{40}$$

The empirical potential of the molecule can be expressed as the summation of the terms in Eqs.(37) and (40) for every bond and every couple of adjacent bonds. The result is the global potential at time t:

$$V(\psi(t)) = \sum_{b \in B_t} V_b(\psi(t)) + \sum_{c \in C_t} V_c(\psi(t)) \tag{41}$$

A recognizable benefit of the above formulation is that the potential energy of a system is separated into individual energy terms according to their physical meaning and to the single atoms. This will be significant also in the atomic-scale finite element approach, which is described later. The harmonic approximation in this context fits the fluctuation of the system around an equilibrium state and it's similar to the linearized theory of elasticity in CM.

Forces are applied on the atoms of the structure. For every atom a, there's a force that can be described by a time-dependent vector

$$f_a(t) = [f_{a,x}(t), f_{a,y}(t), f_{a,z}(t)]^T.$$ (42)

Thus, the external loads acting on the entire molecule are accounted by the sum

$$L(\psi(t)) = \sum_{a \in A_t} (pos_{a[\psi]}(t)^T (f_a(t)).$$ (43)

Finally, the formula for the potential energy and the loading term is given by

$$I(\psi(t)) = V(\psi(t)) - L(\psi(t)).$$ (44)

Some notation must be introduced for the sake of simplicity. Given a bond $b \in B_t$, it's possible to associate to b a time-dependent vector B(t) that satisfies

$$B_b(t) = (pos_{b1}(t)) - (pos_{b2}(t)).$$ (45)

Let $\psi(t)$ and $\Omega(t)$ be two molecular conformations with a one to one relation between their bonds b_ψ and b_Ω. Then it's defined for all $b \in B_t$

$$w_b(\psi(t), \Omega) = \frac{B_{b[\psi]}(t)}{(l_{b[\psi]}(t))^2} (B_{b[\Omega]})$$ (46)

where $B_{b[\psi]}$ is the vector related to the bond in Ω and similarly for $B_{b[\Omega]}$. The lengths $l_b(t)$ are defined as in Eq.(36) and, for a generic bond b, it equals

$$l_b(t) = \|B_b(t)\|_2$$ (47)

At every time t, the derivative of V ($\psi(t)$) [238] with respect to the test molecular conformation (t) is given by the formula

$$V'(\psi(t))(\Omega) = \sum_{b \in B_t} V'_b(\psi(t))(\Omega) + \sum_{c \in C_t} V'_c(\psi(t))(\Omega).$$ (48)

whose terms can be expanded as follows

$$V'_{bond}(\psi(t))(\Omega) = \sum_{b \in B_t} k_1 (l_{b[\Omega]} - a_{(Cc)}) \frac{B_{b[\psi]}^T}{l_{b[\varnothing]}} B_{b[\Omega]}$$ (49)

and

$$V'_{angle}\left(\psi(t)\right)(\Omega) = k_p \overline{h}\left(p_{c[\psi]}\right)\left(\frac{B_{c1[\psi]}B_{c2[\psi]}}{l_{c1[\psi]}l_{c2[\psi]}}\right)^T \left(w_{c2}\left(\psi(t)\right)(\Omega) - w_{c1}\left(\psi(t)\right)(\Omega)\right) \quad (50)$$

The function h: $[-1, 1] \to R$ is defined as

$$\overline{h}(x) = h(x)\frac{dh(x)}{dt} \simeq -\left(acos(x) - \frac{2\pi}{3}\right)\frac{1}{\sqrt{1-x^2}}. \quad (51)$$

Finally, an explicit formula for the derivative [238] of the energy sum I of Eq. (44) is obtained:

$$I'\left(\psi(t)\right)(\Omega) = V'_{bond}\left(\psi(t)\right)(\Omega) - L'_{bond}\left(\psi(t)\right)(\Omega), \quad (52)$$

where the derivative of the second term is given by

$$L'\left(\psi(t)\right)(\Omega) = \sum_{a/A_t} pos^T{}_{a[\psi]}\left(f_a(t)\right). \quad (53)$$

From this significant result, it's possible to directly calculate the gradient of I of Eq. (44). More precisely, one has that the i-th component of the gradient of I is given by

$$[\nabla I\left(\psi(t)\right)]_i = I'\left(\psi(t)\right)(\Omega_i)(5.54)$$

$$\Omega_i = [0,0,\ldots,0,1,0,\ldots,0]^T$$

\that has zero elements everywhere except at its i-th position.

The molecular system with $N = |A_t|$ atoms have 6N degrees of freedom: the dynamical simulation computes the trajectory in a 6N phase space. These degrees are represented by two vectors, each one with 3N elements: 3N positions and 3N momenta. The first time-dependent vector is the single molecular conformation time-dependent vector $\forall t \in [t_0, +\infty)$ $\psi(t) \in R^{3N}$ that contains, in a compact form, the three coordinates of every atom in the system. The second is the velocity vector. Then, from Newtonian mechanics, the result is the following dynamical system:

$$\begin{cases} \dfrac{d\psi(t)}{dt} = \Gamma(t) \\[2ex] \dfrac{d\Gamma(t)}{dt} = \dfrac{\nabla V\big(\psi(t)\big)}{Nm_a}, \end{cases} \tag{55}$$

where the acceleration of the system is calculated as the force given by the negative of the gradient of the potential energy plus the applied forces and all is divided by the mass of the entire molecule. Initial conditions for this dynamical system are the molecular conformation $\psi(t_0)$ and the initial velocities of the particles (t_0). The atomic mass of carbon is 12.0107 g • mol^{-1} and, hence, the mass of a single atom of carbon is

$$m_a \simeq 1.99.10^{-26}\,Kg,$$

recalling the standard conversion between atomic mass and SI units.

Forces, as usual in that kind of modeling, are obtained as the negative gradient of the scalar potential, which depends on the relative positions of the atoms.

The specified formulation of the forces is

$$F(t) = -\nabla V\big(\psi(t)\big). \tag{56}$$

When the loads are constant over time, one has that

$$\forall t \in [t_0, +\infty)\, \forall a \in A, f_a(t) = f_a \in R^3 \tag{57}$$

and when that condition holds, the presence of a conservation law of the total energy is implied by the formulation. In this case, the total energy E of the system is conserved and can be written as

$$E\big(\psi(t)\big) = I\big(\psi(t)\big) + k\big(\Gamma(t)\big) = E. \tag{58}$$

In Eq. (58), K(Γ(t)) is the kinetic energy and it is given by

$$K\big(\Gamma(t)\big) = \frac{1}{2}m_a \sum_{a/A_t}\big(vel_a(t)^T\, vel_a(t)\big), \tag{59}$$

where vela is the velocity function of the atom $a \in A$.

5.2   ATOMIC-SCALE FINITE ELEMENT METHOD

The Atomic-Scale Finite Element Method [261] (ASFEM) has been recently proposed as an efficient alternative to MM with comparable accuracy. Moreover, it is suitable for a number of approaches for large-scale static problems. The ASFEM has an approximately linear complexity on the number of atoms and, hence, can be applied to larger non-linear systems.

5.2.1   ATOMIC-SCALE FINITE ELEMENTS

A state of ground energy corresponds to the equilibrium configuration of a solid. In standard FEM, a continuous solid is partitioned into a finite number of elements, each one with its set of discrete nodes. The energy minimization of the solid is obtained by the appropriate determination of the molecular conformation. Likewise, in MM the calculation of the atom positions is based on a similar energy minimization. Let a molecular system have N atoms, and then the energy stored in the atomic bonds is denoted by:

$$V_{tot}\left(\left[pos_1; pos_2;...; pos_N\right]\right),\tag{60}$$

where $[pos_1; pos_2;...; pos_N] = \psi$ is a representation of the conformation vector.

$$V_{tot}\left(\left[pos_1; pos_2;...; pos_N\right]\right)=\sum_{i=1}^{N}\sum_{j=i+1}^{N}V_p\left(pos_j - pos_i\right),\tag{61}$$

where$V_p(r)$ is a potential function for bend stretch. Alternatively, also non-binary potentials can be considered when angles between bonds are involved. The total energy of the system is

$$E_{tot}\left(\psi\right)=V_{tot}\left(\psi\right)-\sum_{i=1}^{N}\overline{F}_i.\psi_i,\tag{62}$$

with the external force F applied to the i-th atom, as described in the previous Section. The state of minimal energy has

$$\frac{\partial E_{tot}}{\partial \psi}=0\tag{63}$$

Let $\psi^{(0)}$ be an initial guess of equilibrium state, then the Taylor expansion of E_{tot} around $\psi^{(0)}$ yields the approximation

$$E_{tot} \simeq E_{tot}\left(\psi^{(0)}\right)$$

$$+\frac{\partial E_{tot}}{\partial \psi}|\psi=\psi^0.\left(\psi-\psi^{(0)}\right)$$

$$+\frac{1}{2}(\psi-\psi^{(0)})^T.\frac{\partial^2 E_{tot}}{\partial \psi}|\psi=\psi^0.\left(\psi-\psi^{(0)}\right). \tag{64}$$

The combination of Eq. (63) with Eq. (64), yields the equation

$$K(\psi)u = P(\psi), \tag{65}$$

where $u = \psi-\psi^{(0)}$ is the displacement, K is the stiffness matrix

$$K = \frac{\partial^2 E_{tot}}{\partial \psi \partial \psi}\bigg|\psi=\psi^0 = \frac{\partial^2 V_{tot}}{\partial \psi \partial \psi}\bigg|\psi=\psi^0 \tag{66}$$

and P is the non-equilibrium force vector, given by

$$P = \frac{\partial E_{tot}}{\partial \psi}\bigg|\psi=\psi^0 = \overline{F}-\frac{\partial V_{tot}}{\partial \psi}\bigg|\psi=\psi^0, \tag{67}$$

where F is the force vector seen in Eq.(62). When no bifurcations are present, the resulting non-linear system can be solved with iterative methods [261] and it is solved iteratively until P reaches zero. For atomistic interactions with pair potentials, K and P can be obtained from the continuous model as a representation of non-linear spring elements. The ASFEM can account for multi body inter atomic potentials that are significant interactions in the molecular models where the bond angles are considered (Fig. 5.6).

Fig. 5.6. Representation of the i-th finite element for an atom chain.

5.3   APPLICATION TO ATOM CHAINS AND NANOTUBES

In the case of a simple one-dimensional chain of N carbon atoms, an example of multibody inter atomic potential for the bond between the (i−1)-th and the i-th atoms is given by

$$V_{(i-1,i)} = V\left(x_i - x_{i-1}; x_{i-1} - x_{i-2}\right) + V\left(x_i - x_{i-1}; x_{i+1} - x_i\right) \quad (68)$$

For this inter atomic potential, the first order derivative with respect to the i-th atom is given by

$$\frac{\partial V_{tot}}{\partial x_i} = \frac{\partial \left[V_{i-2,i-1} + V_{(i-1,i)} + V_{(i,i+1)} + V_{(i+1,i+2)}\right]}{\partial x_i} \quad (69)$$

that depends on the first and second order nearest neighbor atoms, as shown by Fig. 5.7. For the i-th element, the element stiffness matrix is given by

$$K_i^{(e)} = \begin{bmatrix} \dfrac{\partial^2 V_{tot}}{\partial x_i \partial x_i} & \dfrac{\partial^2 V_{tot}}{2\partial x_{i-1} \partial x_i} & \dfrac{\partial^2 V_{tot}}{2\partial x_{i-2} \partial x_i} & \dfrac{\partial^2 V_{tot}}{2\partial x_{i+1} \partial x_i} & \dfrac{\partial^2 V_{tot}}{2\partial x_{i+2} \partial x_i} \\[3mm] \dfrac{\partial^2 V_{tot}}{2\partial x_{i-1} \partial x_i} & 0- & 0 & 0- & 0 \\[3mm] \dfrac{\partial^2 V_{tot}}{2\partial x_{i-2} \partial x_i} & 0- & 0 & 0- & 0 \\[3mm] \dfrac{\partial^2 V_{tot}}{2\partial x_{i+1} \partial x_i} & 0- & 0 & 0- & 0 \\[3mm] \dfrac{\partial^2 V_{tot}}{2\partial x_{i+2} \partial x_i} & 0- & 0 & 0- & 0 \end{bmatrix} \quad (70)$$

with reduced versions of the matrix for the atoms near the ends. The non-equilibrium force vector for the element is

$$\begin{bmatrix} \overline{F}_i - \dfrac{\partial V_{tot}}{\partial x_i} \\ 0 \\ 0 \\ 0 \\ 0 \end{bmatrix} \quad (71)$$

and the gradient can be computed explicitly with Eq.(54). Differently from the standard FEM, no interpolation approximation is employed since the positions of the atoms coincide with the positions of the nodes and, therefore, the obtained element stiffness matrix and non-equilibrium stiffness vector are accurate.

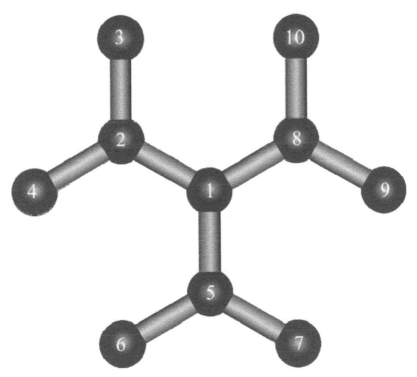

FIGURE 5.7   Representations of a unit cell for a carbon nano tube.

In the application of the ASFEM to carbon nanotubes [261], each carbon atom interacts with both first and second nearest neighbor atoms, since those are the most relevant interactions that must be accounted for [201, 253]. These iterations are the result of the bond angle dependence in the inter atomic potential as represented by Eq. (4). As a result, ten carbon atoms are considered in each element, as represented for the example element show in Fig. 5.7. Since the potential in Eq. (41) has two components, namely the bond and the angle parts, also the stiffness matrix can be decomposed as follows

$$K = K_b + K_a, \tag{72}$$

where the K_b accounts for V_b and K_a for V_a. The simple analytical form of the potential permits the easy computation of the components of K. For a given element centered at the i-th atom, only the relation with other nine atom positions is significant. The position of every atom in R3 is identified by exactly three parameters and, hence, only 30 or less non-zero elements will appear in every row and every column of the stiffness matrix, since the topological distribution follows the honeycomb pattern. The same applies to the element stiffness matrix, which is given by

$$K^{(e)} = \begin{bmatrix} \left[\dfrac{\partial^2 V_{tot}}{\partial \psi_1 \partial \psi_1} \right]_{3\times3} & \left[\dfrac{\partial^2 V_{tot}}{2\partial \psi_1 \partial \psi_i} \right]_{3\times27} \\ \left[\dfrac{\partial^2 V_{tot}}{2\partial \psi_i \partial \psi_1} \right]_{27\times3} & [0]_{27\times27} \end{bmatrix} = K_b^{(e)} + K_a^{(e)}. \tag{73}$$

Figure 5.8 shows the sparsity pattern of K_b, where it's clear that the central atom interacts only with the atoms whose local number is in Ref.[262]. Similarly, Fig. 5.9 shows the sparsity pattern of K_a, showing the relations with the remaining six atoms. Finally, Fig. 5.10 shows the sparsity pattern of K, as defined by Eq.(73). The non-equilibrium force vector for the element is given by

$$P^{(e)} = \begin{bmatrix} \bar{F}_1 - \dfrac{\partial V_{tot}}{\partial \psi_1} \\ [0]_{27\times1} \end{bmatrix}. \tag{74}$$

Hence, because of the regularity of the lattice and of the locality of the interactions, the stiffness matrix K is sparse. This is graphically shown by (Figs. 5.15–517) for an armchair tubule (6, 0), whose local numbering is helicoidally as visualized by (Fig. 5.18). Similarly, the sparsity pattern for the stiffness matrix of a zigzag tubule (25, 25) is shown by (Figs. 5.11–5.13), whose numbering is represented by Fig. 5.14. In brief, the maximum number of non-zero elements of the element stiffness matrix is 30 and, hence, an upper bound to the number of non-zero elements in K is 30N, whereN is the number of atoms in S. For order-N sparse K, the computational cost in time required by the solution ofEq.(65) is also linear [263].

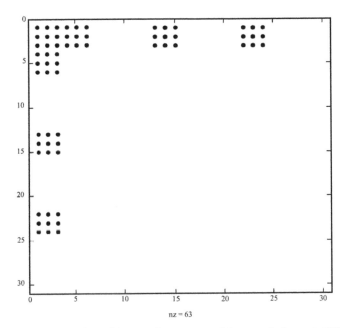

nz = 63

FIGURE 5.8 Visualization of the sparsity patterns of the stretch element stiffness matrix.

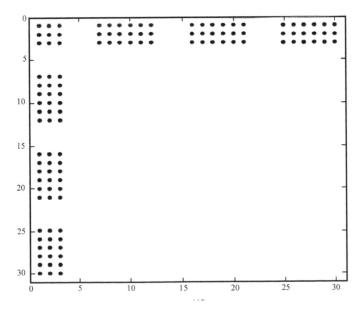

FIGURE 5.9 Visualization of the sparsity patterns of the bend element stiffness matrix.

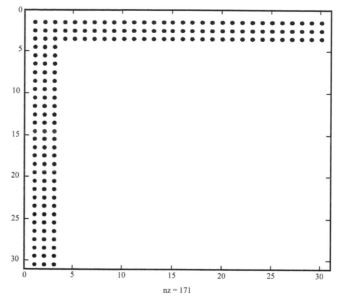

FIGURE 5.10   Visualization of the sparsity patterns of the element stiffness matrix.

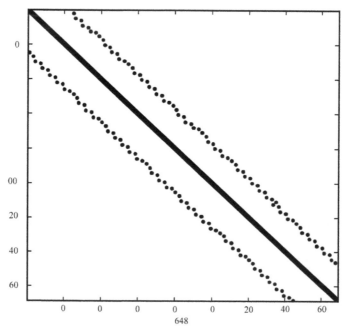

FIGURE 5.11   Visualization of the sparsity pattern of the stretch stiffness matrix for an armchair nano tube (dots are full 3*3 sub matrices).

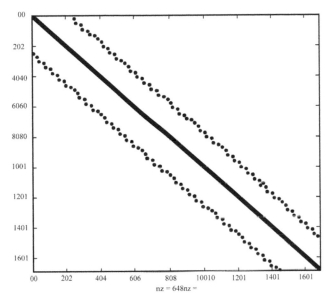

FIGURE 5.12 Visualization of the sparsity pattern of the bend stiffness matrix for an armchair nano tube (dots are full 3*3 sub matrices).

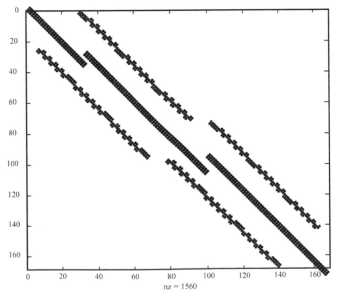

FIGURE 5.13 Visualization of the sparsity pattern of the stiffness matrix for an armchair nano tube (dots are full 3*3 sub matrices)

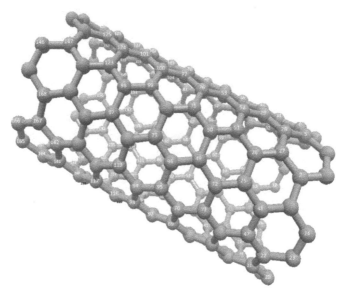

FIGURE 5.14　Global numbering of the atoms in armchair nano tube.

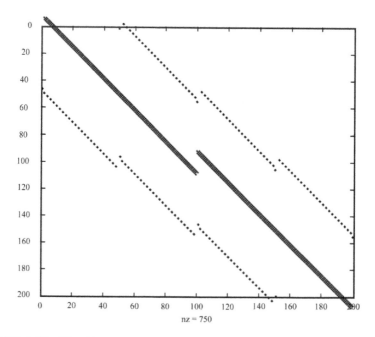

FIGURE 5.15　Visualization of the sparsity pattern of the stretch stiffness matrix for an armchair nano tube (dots are full 3*3 sub matrices).

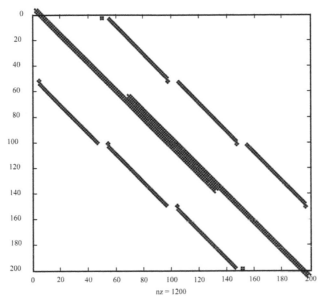

FIGURE 5.16 Visualization of the sparsity pattern of the bend stiffness matrix for an armchair nano tube (dots are full 3*3 sub matrices).

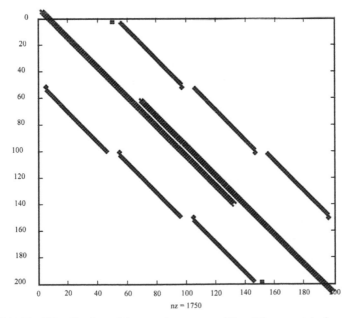

FIGURE 5.17 Visualization of the sparsity pattern of the stiffness matrix for an armchair nano tube (dots are full 3*3 sub matrices).

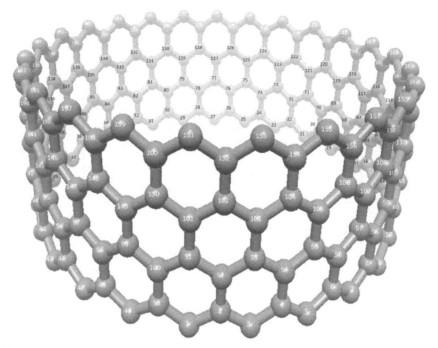

Figure 5.18　Global numbering of the atoms in armchair nano tubes.

In the computer implementation, the stiffness matrices have been stored with the extension of the Compressed Row Storage (CRS) that include only the upper triangular matrix. This is possible because K is also symmetric, being all the atom relations in the mechanics of the lattice symmetric as well.

5.4 NONLINEARITY AND STABILITY

The non-linear resolution of Eq. (65) was obtained by the iteration of the four steps that follow:

1.　Explicitly compute P (ψ (i)) and check its Euclidean norm;
2.　Explicitly compute K (ψ (i));
3.　Solve u (i+1) from (4.65);
4.　Update(i+1);

with i = 0, 1,... η until

$$\left\| P(\psi^{(\eta)}) \right\|_2 \le \epsilon \tag{75}$$

where ε is a predefined tolerance, that in the following computations is set to the value

$$\epsilon = 10^{-6} \tag{76}$$

Every iterations of the computation has a temporal complexity that grows linearly, since the effort to compute K and P is linear and also the third step is linear, because of the sparsity of K. However, since the energy minimum of the non-linear system is reached iteratively, the overall computational effort of the method is O(NM), where M is the number of iteration steps. In general, M is a function of N but, in an immediate neighborhood of ψ°, the Taylor expansion holds. Therefore, with stable atomic structures and harmonic inter atomic potentials, the number of iterations M is approximately independent on N and can be considered a constant value [264].

The selection of the initial guess $\psi^{(\circ)}$ is an important factor for the extraction of an approximately constant value for M. In all the following simulations, the initial guess $\psi^{(\circ)}$ was the parametrically generated SW-CNT at rest. Thus, the initial positions of the atoms are the coordinates generated according to the rules of Section 5.1. As a result, given the tolerance expressed by Eq.(76), the value of M for molecules with 100 <N <2200 (for tubules with arbitrary chirality) was almost constant in the numerical simulations that follow, ranging from 37 to 42.

Stability and convergence of the ASFEM are discussed in Ref. [261]. Both the stability and the convergence of the method are ensured when the energy of the system decreases in every step of the computation.

$$U^T P > 0 \tag{77}$$

That means that and a sufficient condition is that the matrix K is positive definite. This condition holds for problems that don't involve non-linear bifurcations or material softening. The bifurcation strains at about 7% for axial loads agree with MM approaches [12].

In the following simulations, no bifurcation was observed for the measurement of the Young's modulus and the Poisson's ratio. However, bifurcations are present with the simulation of the buckling under heavy axial load, where the tube length was reduced far more than 10% of its initial length at rest. This situation is presented by Fig. 5.19, where a compressive force of –32 nN is applied on a (16, 0) nano tube.

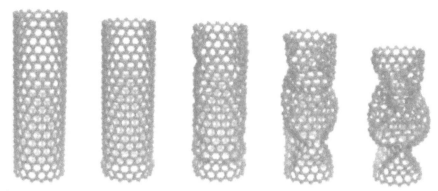

FIGURE 5.19   Five frames (separated by $0.36.10^{-12}$ s) of the buckling simulation of a (16, 0) SWCNT under an axial force F_z=-32 nN.

5.5   DYNAMICS OF THE MOLECULAR SYSTEM

5.5.1. TIME INTEGRATION

The explicit formulation of the gradient permits also the simulation of the dynamics of the system. Being a dynamical system with a direct evaluation of the gradient, the time integration scheme is the core of the dynamical computation. It integrates the equation of motion of the interacting atoms in order to describe their trajectories. Here, the time integration is based on the Verlet algorithm [265], a popular integration method in Molecular Dynamics (MD) [103, 219]. Proceeding to the third order the with the Taylor expansion in t of the conformation (t), it gives

$$\begin{cases} \psi(t+h)=\psi(t)+h\dfrac{d\psi(t)}{dt}+\dfrac{h^2}{2}\dfrac{d^2\psi(t)}{dt^2}+\dfrac{h^3}{6}\dfrac{d^3\psi(t)}{dt^3}+O(h^4) \\ \psi(t-h)=\psi(t)-h\dfrac{d\psi(t)}{dt}+\dfrac{h^2}{2}\dfrac{d^2\psi(t)}{dt^2}-\dfrac{h^3}{6}\dfrac{d^3\psi(t)}{dt^3}+O(h^4) \end{cases} \quad (78)$$

with a uniform time step $h>0$. The sum of the two terms of Eq. (78) gives

$$\psi(t+h)+\psi(t-h)=2\psi(t)+h^2\dfrac{d^2\psi(t)}{dt^2}+O(h^4). \quad (79)$$

and this finally permits

$$\psi(t+h) = \psi(t-h) + 2\psi(t) + h^2 \frac{d^2\psi(t)}{dt^2} + O(h^4) \tag{80}$$

Recalling from Eq.(55) that

$$\frac{d^2\psi(t)}{dt^2} = -\frac{\nabla I(\psi(t))}{Nm_a} \tag{81}$$

one has the integration scheme that follows

$$\psi(t+h) = -\psi(t-h) + 2\psi(t) - h^2 \frac{\nabla I(\psi(t))}{Nm_a} + O(h^4) \tag{82}$$

that is accurate, stable and allows easy implementations. Its truncation error decreases fast when the time step h is reduced, because it is of the order of O (h^4). The third derivative of the molecular conformation and Γ (t) don'tappear and the gradient can be computed as the direct evaluation of the Eq.(52). The Verlet scheme offers time-reversibility and its area preserving, too.

It's important to point out that the local error ε (ψ (t + h)) in position is O(h^4) but its cumulative counterpart $\varepsilon(\psi$ (t + T)) is O(h^2) because, in a given interval of length T = mh, it is proved by induction that

$$\varepsilon(\psi(t+mh)) = \frac{m(m+1)}{2} + O(h^4) \tag{83}$$

$$\varepsilon(\psi(t+T)) = \left(\frac{T^2}{2h^2} + \frac{T}{2h}\right)O(h^4) = O(h^2). \tag{84}$$

Even if the evaluation of Γ (t) is not strictly necessary in the evolution of the system, it is useful to obtain its numerical result in order to periodically check the conservation of the total energy of the system, as seen in Eq.(58); of course, this can be applied only under the conditions of Eq.(57). The potential and kinetic energies flow back and forth during the computation but the global energy must be constant under a specified threshold σ. Under the value σ, small fluctuations are tolerated (usually $\sigma \approx 10^{-5}$ or less). These fluctuations are mostly the result of truncation errors in the time integration and their magnitude can be strongly reduced by making the time step h smaller. In the very long run, small drifts can be experienced and are mostly caused by the excessive size of the time step of the simulation.

To compute the velocities, the simplest method is to proceed at the next step with the central differences. The result is given by

$$\Gamma(t) = \frac{\psi(t+h) - \psi(t-h)}{2h} + O(h^4) \tag{85}$$

But the local error in velocity is $O(h^2)$ and it's two orders less precise than the computed local molecular conformation. Velocities are needed only for the energy check and their errors are not cumulative because the values are recomputed when needed. The velocity Verlet scheme allows the calculation of the velocity with the same accuracy as the local conformation. This method is similar to the standard Verlet but explicitly includes the velocities:

$$\begin{cases} \psi(t+h) = \psi(t) + h\Gamma(t) + \dfrac{h^2}{2Nm_a} - \nabla I\big(\psi(t)\big) + O(h^4) \\[2mm] \Gamma(t+h) = -h\dfrac{\nabla I\big(\psi(t)\big) + \nabla I\big(\psi(t+h)\big)}{2} + O(h^4) \end{cases} \tag{86}$$

An alternative approach to overcome the difficulty of the coherent evaluation of the velocities is given by the application of the leap-frog algorithm.

5.6   NUMERICAL SCHEME AND COMPLEXITY OF THE DYNAMICS

The general structure of the computational method can be as following:

1. Compute V (ψ (t));
2. Compute K(Γ(t));
3. Compute $[\nabla I(\psi(t))]_i = I'(\psi(t))(\Omega_i)$ for all i= 1, 2,...N;
4. Compute ψ (t + h)← ψ(t) + hΓ(t) + (h²/2Nma)− $\nabla I(\psi(t))$;
5. Compute Γ (t + h)← −h($\nabla I(\psi(t))$+$\nabla I(\psi$(t+h)))/2
6. Check the energy conservation

The main loop iterates

$$\frac{T - t_0}{h} = O\left(h^{-1}\right)$$

(87)

where it is independent on the number of atoms $N = |A_t|$. The computation of the gradient requires the direct evaluation of every element of the vector $\nabla I'(\psi(t))$ at every step of the main loop. Every element requires the evaluation of the formula in Eq.(52) with $\Omega = \Omega_i$. This evaluation involves many zero terms, since only one atom is displaced when Ω_i is employed. This allows the significant reduction in complexity since a pre calculated list of the other atoms involved is already obtained at the time of the operation. This structure can be stored in a vector, because of the considerations on the sparsity that follow from the previous Section. The number of non-zero components in the list is constant, more precisely lower or equal to 9 (that is the number of atoms that share at least a couple of bonds with the atom that is modified by Ω_i), as seen when the finite element was first introduced. Therefore, every step of the inner loop for the gradient requires a constant number of elementary operations and must be repeated 3N times. The global complexity of the algorithm is given by $O(Nh^{-1})$ that is linear on the number of atoms. Of course, the pre calculation of the atoms involved in every operation can be computed in linear time. Once the Y -cell structure is defined, every bond and every couple of bonds are read; both the atoms that constitute a bond and all three the atoms that constitute a couple of bonds are added (if they are not already present) to the involvement list of all the other atoms that appear in the same bond or couple of bonds. Every step can be done in constant time, since the Y-cell has a constant number of elements, and the global complexity is linear $O(N)$ on the number of atoms. The dynamical scheme permits the simulation of the bending of a (16, 0) nano tube, whose frames are presented in Fig. 5.20. The visualization presents a rippling effect that is similar to the one observed in other computations [266].

Three frames of a torsion action are represented in Fig. 5.21, with torsion of π around the tube axis. The visualization shows the same three-fold symmetry that was simulated by the ECBR [239]. In these computations, the length of the time step is

$$h = 3.10^{-15} s$$

(88)

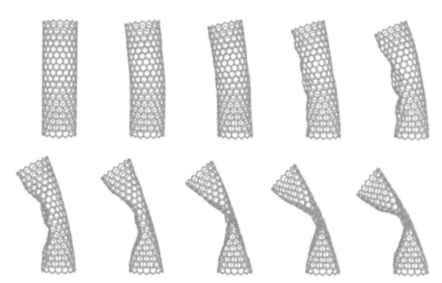

FIGURE 5.20   Ten frame of the bending simulation of a (16, 0) SWCNT.

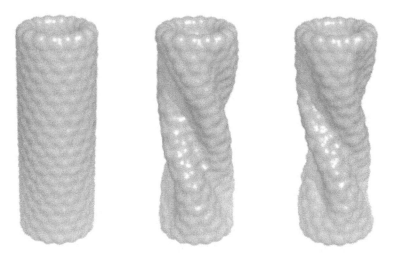

FIGURE 5.21   Three steps of the simulated behavior under torsion of a (16, 0) SWCNT.

5.7   NUMERICAL RESULTS

The obtained results permit the analysis of the dependence of length, diameter and elasticity on curvature, helicity and presence of defects in the molecules. The computed results are presented, compared and validated with the available experimental and computational data obtained so far. Moreover, this section discusses the assumptions for the continuous reference model and their applicability.

5.8   DIAMETERS AND LENGTHS AT THE ENERGY GROUND

The simulation of the length at rest of several SWCNTs, with the same number of Y-shaped elements in the axial direction, shows that the curvature influences the length at rest of the nano tube: the smaller the diameter, the larger the discrepancy between the length obtained from the rolling-up of a graphene lattice and the effective length of the tubule at its energy ground. These results can be plotted as a function of the diameter for all the armchair tubules (n, 0) with n ∈ {3, 4,...17}. Similarly, the results for (n, n) zigzag SWCNTs are presented in Fig. 5.22 for all the n ∈ {5, 6,...,30}. All the generated models are three hexagons long. The results are compared in the combined plot of Fig. 5.23. In these plots, the length discrepancy is defined as

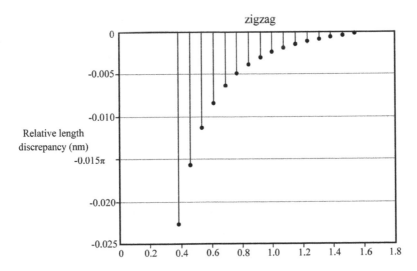

FIGURE 5.22   Plots of the length discrepancy at the energy ground for zig-zag nanotubes.

where l_0 is the length on a graphene sheet approximation and l_g is the length at the energy ground for the tube. As a consequence of these results, the smaller the radius of the tube, the less the hexagons are stretched in the axial direction. The length discrepancy is reduced with the increase of the diameter and follows the approximation

$$\delta(d) \propto \frac{1}{d^2} \tag{89}$$

for all the tubules, almost vanishing for diameters smaller than 1.2 nm. This agrees with the fact that the local effects of the curvature decrease for larger diameters. Moreover, it's also observed that the length discrepancy has a larger absolute value for zigzag nanotubes. In fact, Fig. 5.23 shows that the length discrepancy is more than double for zigzag nanotubes with the diameter

$$d \in [0,4,1]$$

Figure 5.24 shows a combination of the armchair tubes used in the simulations, showing the different diameters and the global numbering. As seen for the ground energy lengths, the curvature of the cylindrical surface strongly influences the lattice configuration of narrow tubes. Another effect of the curvature is the discrepancy between the estimation of the diameter with the norm of the chirality vector and the real diameter at the energy ground. Figure 5.23 shows the discrepancy between the estimated radius from graphene and the energy ground radius for armchair nanotubes. Similarly, Fig. 5.24 shows the discrepancy for zigzag nanotubes and, finally, the results are compared in Fig. 5.25.

The radial discrepancy δ_r is defined as

$$r = d_t - d_g, \tag{90}$$

where $d_t \pi = \|C_h\|_2$ and d_g is the diameter at the energy ground. Hence, the smaller the radius, the larger d_g becomes with respect to d_t. This effect and the length discrepancy have the same cause that is the fact that planar Y – shaped cells become tetrahedral structures because of the curvature and the $2\pi/3$ equilibrium angles are not respected, increasing the bending potential energy. The comparison of the plots shows that zigzag nanotubes exhibit larger effects, as it was also observed for the length discrepancy. Again, the effects of curvature locally vanish with larger diameters. Figure 5.26 shows the differences between the initially computed model and its energy ground configuration for a zigzag (5, 5) nano tube.

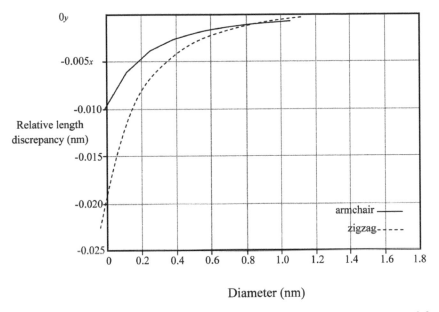

Diameter (nm)

FIGURE 5.23   Plot comparisons of the length discrepancies at the energy ground for nanotubes.

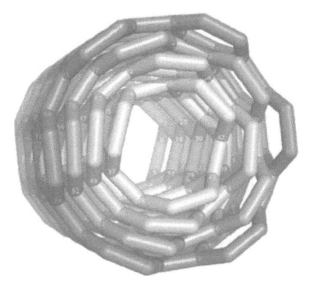

FIGURE 5.24   Nested visualizations of the armchair nano tube for the simulation of the length discrepancies, showing the diameters and their separate global numbering.

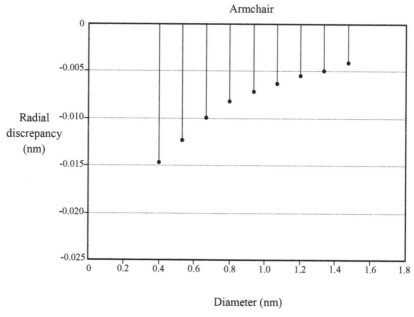

FIGURE 5.25   Plot of the radial discrepancy for armchair nanotubes.

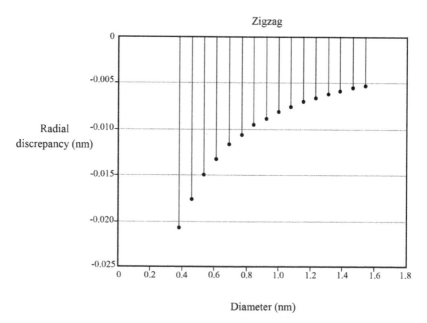

Figure 5.26   Plot of the radial discrepancy for zig-zag nanotubes.

In brief, zigzag nanotubes present a double intensity of the length discrepancy, when compared with armchair tubes. Similarly, the former class presents a radial discrepancy whose intensity is approximately 4/3 of the value obtained for the latter. As a result, SWCNTs can be conveniently described as rolled graphene sheets, but it's important to point out that this is not their ground energy configuration, since curvature is the cause of the discrepancy effects both in length and in diameter. Hence, the results show that the minimized configurations are shorter and larger, which is a result that is comparable with existing investigations [267, 268]. In order to obtain more accurate results, additional specific terms for the potential energy should be considered (Figs. 5.27 and 5.28).

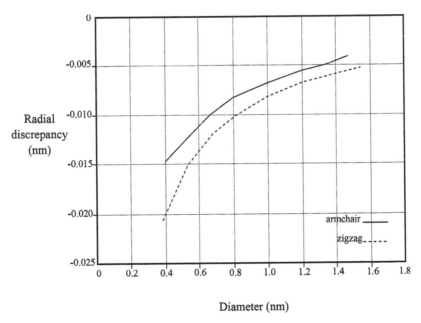

Diameter (nm)

FIGURE 5.27   Plot comparison of the radial discrepancies for nanotubes.

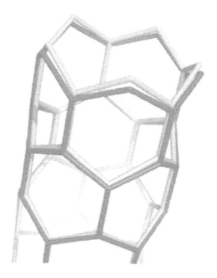

FIGURE 5.28   Comparative visualization of the generated model (white) of a (5, 5) zigzag nano tube and its energy-minimized configuration (black).

5.9   YOUNG'S MODULI

The first step to describe a novel structural material is to present its Young's modulus E. This parameter is related to the chemical bonding of the atoms that constitute the specific material. Recalling the definition for a thin rod of isotropic and homogeneous material of cross-sectional area A_0 and of length l_g, E is given by

$$\bar{E} = \frac{F}{A_0} \frac{l_g}{\Delta l},$$ (91)

where Δl is the difference in length after the load and F is the force applied to the cross-sectional area. A possible convention is

$$A_0 = \left(\frac{d_g}{2}\right)^2 \pi.$$ (92)

and this choice will be discussed later. Figure 5.29 shows the obtained results of the Young's moduli for armchair nanotubes as a function of their diameters.

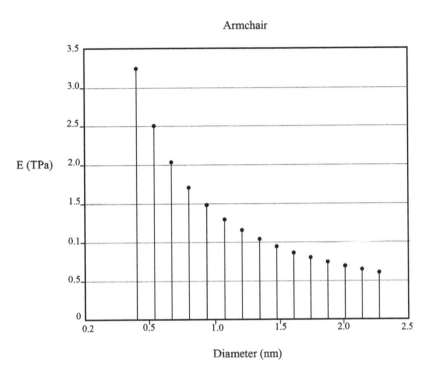

FIGURE 5.29   Plots of the young's moduli for armchair nanotubes.

Likewise, Fig. 5.30 plots the results for zigzag nanotubes and, finally, Fig. 5.31 compares the results. In the simulations, models with comparable length have been used, with an applied force of −0.05 nN per atom in the axial direction of the tubule. The force was applied only on one surface and no bifurcation occurred.

The elastic properties of the SWCNTs have been analyzed by several authors through CM and the recourse to concepts of continuum theory is frequent even when the continuous formulation is not explicit. Nevertheless, it was noted [12] that the relevance of a continuous assumption for co-valent-bonded systems of only a few atoms in diameter is far from obvious.

The continuous idealization of the elastic behavior of a SWCNT is often approached in two ways, both with the basic assumption of an equivalent linearly elastic and isotropic material. SWCNTs can be seen as:

- thin shells; and
- full beams.

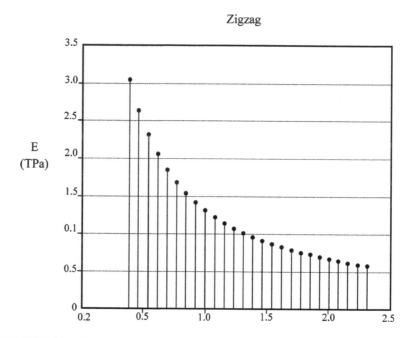

FIGURE 5.30   Plot of the young's moduli for zig-zag nanotubes.

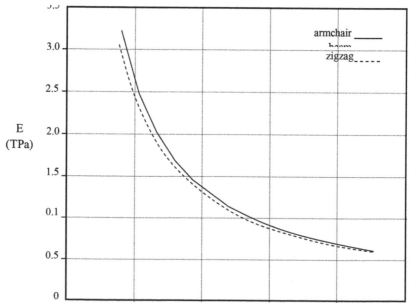

Figure 5.31   Plot comparison of the young's moduli for nanotubes.

In the first case [12, 18, 51, 63, 72], the definition of a wall thickness is required for the definition of the model. The most common conventional thickness is the value of the interlayer distance of graphene sheets in graphite. That parameter is d_p. In this approach, the tubules occupy the volume of a shell that is one atom thick, since their electronic clouds occupy a certain volume of space. However, thin shell theories seem to be inadequate for two reasons. First, the selection of the interlayer thickness with reference to the elastic thin shell leads to inconsistencies between the axial and the bending moduli of nanotubes [12]. Second, every other selection of the thickness is arbitrary. In fact, the fictitious thickness with the value of 0.066 nm [51] presents consistency between the elastic moduli but is not representative of any physical feature of the molecule.

In the previous description, the elastic behavior has been modeled with the full beam approach. More precisely, the tubules have been modeled as cylindrical beams with diameter given by the one of the molecule at its ground energy configuration, as done by [269] with Tight Binding (TB). This differs from most of the similar full beam models, where the considered diameter is computed according to the value of the estimation d_t [270, 271]. Of course, these two conventions tend to coincide for large diameters, but the differences can be significant for small diameters, because of the radial discrepancies described before. It's important to point out that the variation in Young's modulus depends on the square of the variation in the considered diameter, since

$$\bar{E}(d) = \frac{F}{(d/2)^2} \frac{\lg}{\Delta l} \propto \frac{1}{d^2}. \tag{93}$$

In order to compare the obtained results with thin shell data, the Young's moduli for the thin shell model are converted with the new area A_t, which is defined as follows:

$$A_t = \left[\left(\frac{d_t}{2} + \frac{d_p}{2} \right)^2 - \left(\frac{d_t}{2} - \frac{d_p}{2} \right)^2 \right] \pi \tag{94}$$

The obtained thin shell moduli are presented in Fig 5.32. From the plotted data it's clear that the Young's modulus tends, when the diameter tends to infinity, to the same Young's modulus of graphene (Fig. 5.33).

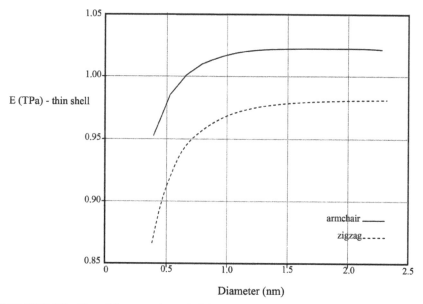

FIGURE 5.32   Plot of the young's moduli for armchair and zigzag nanotubes as a function of their diameters under the thin shell assumption.

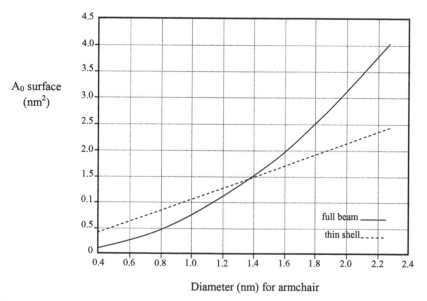

Figure 5.33   Plot of the surface for armchair nanotubes.

For tubes with small diameters, the thin shell model gives a decreased Young's modulus, while the full beam approach yields a larger modulus. This is due to the fact that the discrepancy between the reference surfaces is large for small radii and, thus, the validity of the continuum assumption is controversial in these conditions. Various thicknesses yield different values for the Young's modulus, and the used values range from 0.066 nm [51] to 0.34 nm, the latter being the interlayer distance dp. Figure 5.34 plots the different surface values for A_o, comparing the thin shell with the full beam models for armchair nanotubes. Figure 5.34 plots the surfaces, as a function of the diameter, for zigzag nanotubes. The resulting plots show that the surfaces almost coincide for d≈1.4 nm. This is due to the fact that Eqs. (92) and (94) yield the same value for the specified diameter and, hence, the full beam and the thin shell present the same results in this situation. For that surface and that diameter, the obtained Young's modulus is 1.03 Tpa for armchair tubes and 0.97 TPa for zigzag tubes.

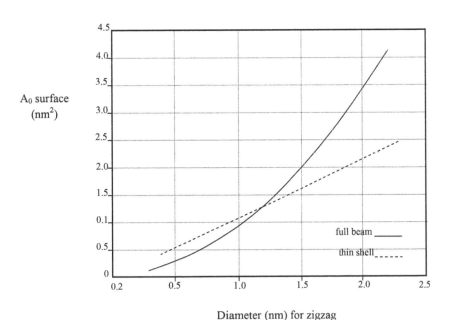

Diameter (nm) for zigzag

FIGURE 5.34   Plot of the surface for zigzag nanotubes.

In general, the experimentally and computationally predicted Young's moduli for SWCNTs range from 0.5 to 5.5 TPa [11, 12, 47, 52, 272]. The large scattering is not only due to the gap between computational and experimental approaches. From the computational point of view, several factors that may strongly influence the results can be summarized as follows:

- different methods (MM, TB, CM, QCM),
- different molecular parameters (specific potentials, fitting of experimental coefficients),
- different reference models (thin shell, equilibrium beam),
- different computational conditions (load applied to the axis, load applied on a single side, load as a stretching force).
- From the experimental point of view, the factors that may diffuse the results can be summarized as:
- different synthesis of the sample (CVD, laser ablation),
- different measurement techniques (loading with AFM, resonance, Raman scattering),
- presence of defects and their type (SWD, LVD),
- challenging identification of the sample (chirality, length, diameter).

The obtained results $E_a = 1.02$ TPa and $E_z = 0.98$ TPa, when normalized with the same thin shell model, show agreement with the data obtained so far. Regardless of the reference model, the simulations suggest that armchair nanotubes are slightly stiffer than zigzag tubes, but the relative difference is upper bounded by 5% and, thus, can be considered negligible, as noted by several authors [250, 268] with a wide range of methods. In brief, the Young's modulus approximately doesn't dependent on chirality. It always strongly depends on diameter for small radii but, for large radii, depends only with the full beam model. Not surprisingly, the Young's modulus for tubes with large diameters tends to the value of the Young's modulus of graphene [216, 269].

5.10   POISSON'S RATIOS

The Poisson's ratio v is a measure of the tendency of a material, when stretched in one direction, to get thinner in the remaining directions. More precisely, it's the ratio of the relative contraction strain normal to the applied load and the relative extension strain. In the case of CNTs, assuming that the equivalent material is compressed along the z-axis that coincides with the tube axis, one has

$$v = -\frac{\Delta l}{\lg}\frac{d_g}{\Delta d} = -\frac{\epsilon_z}{\epsilon_x} = -\frac{\epsilon_z}{\epsilon_y}, \tag{95}$$

where Δl is the difference in length after the load and Δd is the difference in diameter after the load. In the numerical simulations, an axial force of -0.05 nN per atom on one surface is used and the minimization doesn't suffer bifurcations. Figure 5.35 plots the Poisson's ratio for armchair and zigzag nanotubes.

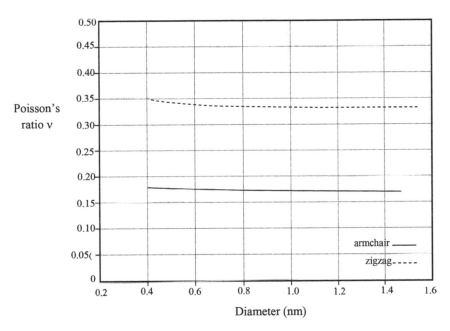

FIGURE 5.35 Plot of the Poisson's ratios for armchair and zigzag nanotubes.

The obtained Poisson's ratios are approximately not dependent on the radius of the tubules. In fact, narrow tubes have slightly larger Poisson ratios that, however, is Upper bounded by 5%. Vice versa, the Poisson's ratio seems to depend in a significant way on the chirality, as observed by other authors [273]. The Poisson's ratios for armchair tubes is $v_a = 0.18$, which is almost half of the value $v_z = 0.33$ for zigzag tubes. Consequently, it's always necessary to specify the chirality of the tube when the data is analyzed. A number of authors [12, 47, 238, 272, 274] obtained values for the Poisson's ratio that range from 0.15 to 0.34.

5.11   SHEAR MODULI

The shear modulus G, also called the modulus of rigidity, describes the tendency of a material to shear that is to deform its shape at constant volume when acted upon by opposing forces. It is defined as shear stress over shear strain. For an isotropic elastic material the Young's modulus E, the Poisson's ratio and the shear modulus G are related as follows:

$$G = \frac{\bar{\bar{E}}}{2(1+v)}.$$

(96)

Hence, under the assumption of an isotropic elastic material, Figure 5.36 plots the results for the obtained shear moduli for armchair and zigzag nanotubes. Due to the explained difficulties with experimental techniques, there're still a small number of reports on the measured values of shear modulus of SWCNTs. However, in order to compare the data with the thin shell models, the obtained results have been converted to the thin shell model with thickness dp and finally plotted in Fig. 5.37. They slightly

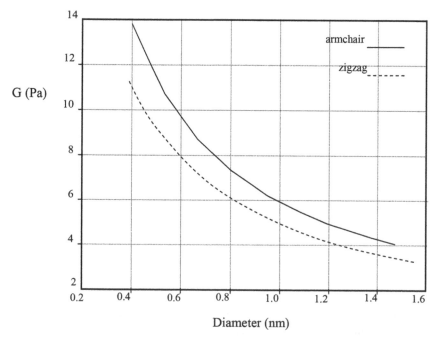

FIGURE 5.36   Plot of the shear moduli for armchair and zigzag nanotubes

depend both on chirality and on diameter, as observed by [32]. Theoretical predictions [47] obtained the average result of 0.5 TPa, which is comparable to the values presented in this thesis and also to that of diamond [183]. The lattice-dynamics model permitted the derivation of an analytical expression for the shear modulus, indicating that the shear modulus is about equal to that of graphene for large radii and smaller for narrow tubes [32]. Moreover, it made possible the observation of a weak dependence of the shear modulus on the chirality of the tubes for small radii.

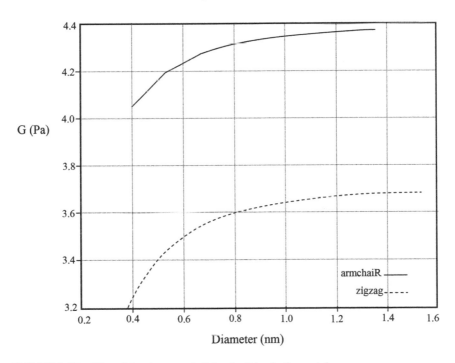

FIGURE 5.37 Plot of the shear moduli for the thin shell nanotubes.

5.12 YOUNG'S MODULI OF DEFECTIVE NANOTUBES1

A number of attempts to resolve the discrepancies from computational and experimental data have concentrated on the role of defects in the elasticity of SWCNTs [40, 275, 276]. The Young's moduli of several defected nanotubes are considered as significant mechanical parameters obtained from the numerical simulation. Models of defective tubules have been created on purpose and theirconfigurations have been optimized with respect to their respective potentialenergies. The minimization process from the initial configuration is shown in Fig. 5.38 for a (8, 0) nano tube with a single-atom vacancy. For the single-atomvacancy, the obtained value of the measured Young's modulus is 1.7714 TPa,which is about 7/8 of the Young's modulus of a defect-free tubule. When twosingle-atom vacancies are present on a (8, 0) tube and are placed on oppositesides with respect to the tube's axis, the computed Young's modulus is as low as1.3358 TPa. Figure 5.39 shows the minimized structure of the (8, 0) tubule with two single-atom vacancies. For a (8, 0) nano tube with the large vacancy defectshown in Fig. 5.40, the measured value of the Young's modulus is 1.5844 TPa.

 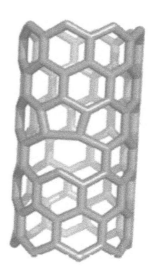

FIGURE 5.38   Parametrically generated model (left) and its ground energy configuration (right) for a (8, 0) nano tube with a single-atom vacancy (only carbon atoms are visualized in the lattice).

The results obtained in this model confirm that the presence of defects implies a reduction in the strength of the nanotubes. More precisely, the presence of a single-atom vacancy yields a Young's modulus that is 7/8 of the modulus of a perfect (8, 0) tubule. Moreover, when the number of single atom vacancies symmetrically doubles, the Young's modulus is approximately reduced to 2/3 of the original value. Larger defects yields

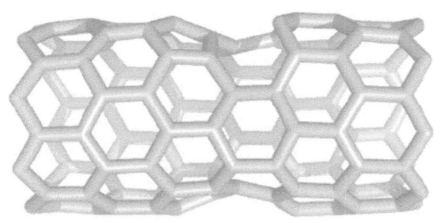

FIGURE 5.39   Minimized configuration of a (8, 0) nano tube with two single-atom vacancies (only carbon atoms are visualized in the lattice).

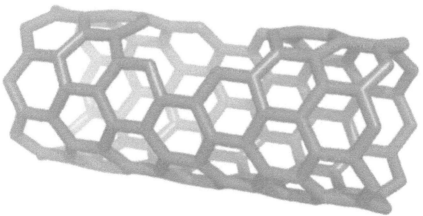

FIGURE 5.40   Minimized configuration of a (8, 0) nano tube with a large vacancy defect (only carbon atoms are visualized in the lattice).

larger decrease in the value of the Young's modulus, as a zero-level LVD reduces the modulus to 3/4 of its initial value. In brief, despite the large reduction of the value of the Young's modulus, SWCNTs exhibit outstanding moduli even when realistic defects are present. This is an excellent property because even high-quality nanotubes are expected to have a significant percentage of defects with the production processes described before.

KEYWORDS

- **Atomic-Scale**
- **Mechanical Behavior**
- **Nanotubes**
- **Nonlinearity and Stability**
- **Numerical Scheme**
- **Numerical Simulation**

MOST IMPORTANT APPLICATIONS OF CARBON NANOTUBES

A1 ELECTROCHEMICAL INTERCALATION OF CARBON NANOTUBES WITH LITHIUM

The basic working mechanism of rechargeable lithium batteries is electrochemical intercalation and de-intercalation of lithium between two working electrodes. Current state-of-art lithium batteries use transition metal oxides (i.e., Li_xCoO_2 or $Li_xMn_2O_4$) as the cathodes and carbon materials (graphite or disordered carbon) as the anodes. It is desirable to have batteries with a high-energy capacity, fast charging time and long cycle time.

It has been considered that a higher Li capacity may be obtained in carbon nanotubes if all the interstitial sites (intershell van der Waals spaces, intertube channels, and inner cores) are accessible for Li intercalation. Electrochemical intercalation of MWNTs [277–278] and SWNTs [279–280] has been investigated by several groups. Figure A1 (top) shows representative electrochemical intercalation data collected from an arc-discharge-grown MWNT sample using an electrochemical cell with a carbon nanotube film and a lithium foil as the two working electrodes.

The high capacity and high-rate performance warrant further studies on the potential of using carbon nanotubes as battery electrodes. The large observed voltage hysteresis is undesirable for battery application. It is at least partially related to the kinetics of the intercalation reaction and can potentially be reduced/eliminated by processing, that is, cutting the nanotubes to short segments.

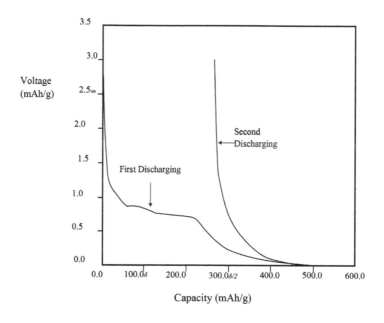

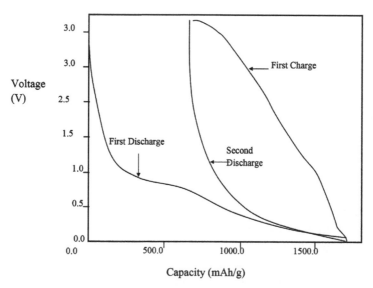

FIGURE A1   Top:Electrochemical intercalation of MWNTs with lithium. Data were collected using 50 mA/h current. The electrolyte was LiClO$_4$ in ethylene carbonate/dimethyl carbonate. **Bottom:**Charge-discharge data of purified and processed SWNTs. The reversible capacity of this material is 1000 mAh/g (using data from Gaoet al. [281]).

A2 CARBON NANOTUBES IN NANOPROBES AND SENSORS

The small and uniform dimensions of the nanotubes produce some interesting applications. With extremely small sizes, high conductivity, high mechanical strength and flexibility (ability to easily bend elastically), nanotubes may ultimately become indispensable in their use as nanoprobes. One could think of such probes as being used in a variety of applications, such as high-resolution imaging, nano-lithography, nanoelectrodes, drug delivery, sensors and field emitters. The possibility of nanotube-based field emitting devices has been already discussed. Use of a single MWNT attached to the end of a scanning probe microscope tip for imaging has already been demonstrated (Fig. A2) [282–283].

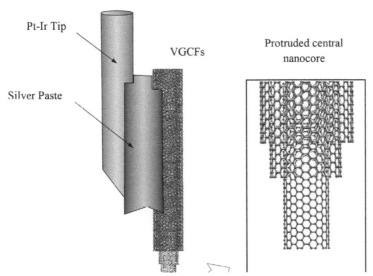

FIGURE A2 Use of a MWNT as an AFM tip. At the center of the Vapor Grown Carbon Fiber (VGCF) is a MWNT,which forms the tip. The VGCF provides a convenient and robust technique for mounting the MWNT probe for use in a scanning probe instrument.

Since MWNT tips are conducting, they can be used in STM, AFM instruments as well as other scanning probe instruments, such as an electrostatic force microscope. The advantage of the nanotube tip is its slenderness and the possibility to image features (such as very small, deep surface cracks), which are almost impossible to probe using the larger, blunter etched Si or metal tips. In addition, due to the high elasticity of the nanotubes, the tips do not suffer from crashes on contact with the substrates. Any impact will cause buckling of the nanotube, which generally is reversible on retraction of the tip from the substrate.

Attaching individual nanotubes to the conventional tips of scanning probe microscopes has been the real challenge. Bundles of nanotubes are typically pasted on to AFM tips and the ends are cleaved to expose individual nanotubes (Fig. A2).

A3 CARBON NANOTUBES IN TEMPLATES

Since nanotubes have relatively straight and narrow channels in their cores, it was speculated from the beginning that it might be possible to fill these cavities with foreign materials to fabricate one-dimensional nanowires. Early calculations suggested that strong capillary forces exist in nanotubes, strong enough to hold gasses and fluids inside them [284]. Thus, nanotubes have been used as templates to create nanowires of various compositions and structures (Fig. A3).

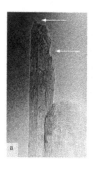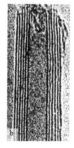

FIGURE A3   Results that show the use of nanotubes as templates. The *left-hand* figure is a schematic that shows the filling of the empty one-dimensional hollow core of nanotubes with foreign substances. (a) Shows a high-resolution TEM image of a tube tip that has been attacked by oxidation; the preferential attack begins at locations where pentagonal defects were originally present (*arrows*) and serves to open the tube. (b) TEM image that shows a MWNT that has been completely opened by oxidation. (c) TEM image of a MWNT with its cavity filled uniformly with lead oxide. The filling was achieved by capillarity.

The critical issue in the filling of nanotubes is the wetting characteristics of nanotubes, which seem to be quite different from that of planar graphite, because of the curvature of the tubes. Wetting of low melting alloys and solvents occurs quite readily in the internal high curvature pores of MWNTs and SWNTs. In the latter, since the pore sizes are very small, filling is more difficult and can be done only for a selected few compounds. It is intriguing that one could create one-dimensional nanostructures by using the internal one-dimensional cavities of nanotubes. Liquids such as organic solvents wet nanotubes easily and it has been proposed that interesting chemical reactions could be performed inside nanotube cavities [285].

A4   CARBON NANOTUBES AS ADSORBENTS FOR REMOVAL OF POLLUTANTS

Adsorption is a simple and efficient method for the removal of organic and inorganic compounds in drinking water treatment. Among the various adsorbents, such as activated carbons (ACs), zeolites, and resins, ACs are one of the most widely used type of adsorbents in water treatment, because of their several merits: broad-spectrum removal capability toward pollutants, chemical inertness, and thermal stability. However, the application of ACs in water treatment also suffers from several bottlenecks, such as slow adsorption kinetics and difficulty for regeneration [286–287].

To overcome the above problems, activated carbon fibers (ACFs) were developed as the second generation of carbonaceous adsorbents. The pores in ACFs are directly opening on the surface of carbon matrix, which shortens the diffusion distance of pollutants to adsorption sites. As a result, ACFs usually possess higher adsorption kinetics than ACs. CNTs, with one-dimensional structure, like miniaturized ACFs. All adsorption sites locate on the inner and outer layer surface of CNTs. With the hollow and layered structures and tunable surface chemistry, theoretically,

CNTs may be a promising third generation of carbonaceous adsorbents because of various physical structures and surface chemistry.

In most cases, several driving forces act simultaneously, including hydrophobic effect, π–π interaction π–π electron-donor-acceptor (EDA) interaction, electrostatic interaction, and hydrogen bonding. Because of the hydrophobic nature of their outer surfaces, CNTs have a strong affinity to organic chemicals, especially to non-polar organic compounds, such as naphthalene [288–289], phenanthrene, and pyrene [290]. Meanwhile, the abundantπ electrons on CNT surfaces enable a strong π–π coupling of aromatic pollutants with the CNT surface. Within the groups of nitroaromatics, the adsorption affinity increased with the number of nitro groups. In the light of the above mechanisms, the morphology of CNTs including nanoscale curvature and chirality of graphene layers is expected to have a great influence on the adsorption of organic pollutants, especially for those with π–π stacking as the interaction force. Gotovac et al. [289] observed remarkable difference between the adsorption capacities of tetracene and phenanthrene on the tube surface of CNTs and difference in their aggregation tendency because of the nanoscale curvature effect and their morphology differences.

The aggregation tendency reduces with increased number of walls, or in other words, reduced nanocurvature. Generally, the aggregation of CNTs follows such an order: single-walled CNTs (SWCNTs)> double-walled CNTs (DWCNTs)>multi-walled CNTs (MWCNTs)(the interlayer spacing between the coaxial layers of MWCNTs was not available for adsorption). As shown in Fig. A4, SWCNTs usually exist as bundles or ropes while MWCNTs are randomly entangled as individual tubes. As a consequence of aggregation, the available outer surface was reduced while new adsorption sites appeared as interstitial channels and grooves between the tubes in CNT boundless.

Zhang et al. [291]focused on numerical study of CNTs to calculate the changes in pore volume and specific surface area caused by aggregation and found that aggregation of CNTs was unfavorable for the adsorption of several synthetic organic compounds (SOCs) on CNTs, since the surface area was more important than the pore volume in adsorption of SOCs. Ultra sonication significantly enhanced the adsorption kinetics of SOCs on CNTs, indicating that the dispersion status of CNTs affected the adsorption kinetics [292].

Surface chemistry is another important factor influencing the CNTs adsorption behavior. Functional groups such as –OH, –C=O and –COOH could be intentionally introduced onto CNT surfaces by acid oxidation or air oxidation. Those functional groups make CNTs more hydrophilic and suitable for the adsorption of relatively low molecular weight and polar contaminants, such as phenol [293] and 1,2-dichlorobenzene [294].

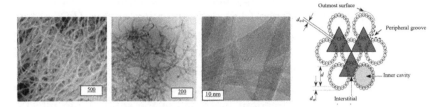

FIGURE A4   SEM images of SWCNT, SEM image of MWCNT, TEM image of SWCNT, and schematic illustration of SWCNT bundle (from left to right).

Figure A5 is an overview of the surface modification and effects of functional groups on the adsorption properties of CNTs. The adsorption of heavy metals on CNTs depends mainly on the specific complexation between metal ions and the hydrophilic functional groups of CNTs. Therefore, without doubt, surface functionalization of CNTs is favorable for the uptake of metal ions. Scientists reported that surface oxidation of CNTs enhanced the adsorption of zinc and cadmium ions from aqueous solutions. Using twosite Langmuir adsorption modeling, they found that the carboxyl-carbon sites of CNTs were over 20 times more energetic for Zn (II) uptake than the unoxidized carbon sites. Another direct consequence of surface modification of CNTs with hydrophilic groups is the improvement of CNT dispersion in aqueous media. Rosenzweig et al. [295] reported that the alcohol (OH) and acid (COOH) moieties on CNTs can determine the aggregation state and accessible sites for copper adsorption.

The surface functionalized CNTs had higher adsorption capacity for copper than pristine CNTs. It is worth noting that increasing the oxygen-containing functional groups is a double-edged sword. It may have an adverse effect on the adsorption of highly non-polar chemicals like naphthalene [296]. Wu et al. [297] systematically investigated the influence of surface oxidation of MWCNT on the adsorption capacity and affinity of organic compounds in water, and found that surface oxidation of MWC-NTs decreased the surface-area-normalized adsorption capacity of organic compounds significantly because of the competition of water molecules; meanwhile, the adsorption affinity of organic chemicals were not altered because of the adsorption interactions (hydrophobic effect π–π. (hydrophobic effect, π–π interaction and hydrogen bond) remained constant.

The complex mechanisms involved in CNT adsorption could be explained from two aspects: thermodynamics and kinetics. Thermodynamic parameters, including free energy change of adsorption (ΔG), enthalpy change (ΔH), entropy change (ΔS) and activation energy (E_a), provide an insight regarding the inherent energetic changes during adsorption. Adsorption of Pb^{2+} [191], trihalomethanes [298], 1-naphthol and phenol [299], methyl orange [300] were demonstrated to be spontaneous processes based on the negative ΔG and the positive ΔS. The positive ΔS values imply that the degree of freedom increased at the solid–liquid interface during the adsorption of contaminants onto CNTs. This could be attributed to the entropy increase of water molecules after the ordered water shells being destroyed in the adsorption process [299].

In terms of adsorption kinetics, the ordered pore structure of CNTs makes it easier for the diffusion of pollutants to adsorption sites [301]. This can be well reflected through the comparison with ACs. ACs is usually rich in micropores, which are sometimes not available for the access of relatively large organic molecules. Ji et al. [302] investigated the adsorption of tetracycline to CNTs, graphite and AC and found that the adsorption affinity of tetracycline decreased in the order of graphite/SWNT > MWNT-AC upon normalization for adsorbent surface area. The weaker adsorption of tetracycline to AC indicated that adsorption affinity was greatly influenced by the accessibility of available adsorption sites. The remarkably strong adsorption of tetracycline to CNTs can be attributed to the strong adsorptive interactions (van der Waals forces, π–π EDA interactions, cation-π bonding) with the graphene surface of CNTs. From a kinetics point of view, Lu et al. [302] studied the adsorption of trihalomethanes to CNTs and a powdered activated carbon (PAC). CNTs reached adsorption equilibrium much faster than the PAC.

This may be explained by the different porous structures of CNTs and PAC. PAC had more micropores in which trihalomethanes have to move from the exterior surface to the inner pores of PAC to reach equilibrium. The more uniform pore structure of CNTs was beneficial for the diffusion of pollutants into the inner pores. Zhang et al. [303]examined the adsorption kinetics of phenanthrene and biphenyl on a granular activated carbon (GAC) and CNTs. Fitting the kinetic data with intraparticle diffusion model indicated that external mass transfer controlled the adsorption of organic compound to CNTs, while intraparticle diffusion dominated in the adsorption of organic compounds onto ACs [303]. Therefore, in well-mixed systems, CNTs are superior to ACs in terms of sorption kinetics. Besides adsorption capacity and kinetics, adsorption selectivity or resistance to harsh environment is another important evaluation criterion for an adsorbent. A number of studies have examined the importance of aqueous chemistry conditions on the adsorption of SOCs by CNTs [293, 304–306]. Effects of solution pH and ionic strength on SOC adsorption by CNTs were somewhat SOC-specific, the extent of which depends on the ionizability and electron-donor-acceptor ability of the involved SOCs [293, 304, 307].

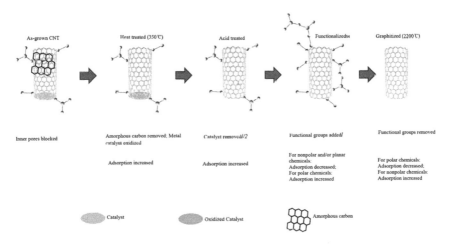

FIGURE A5   Adsorption Properties of CNTs As Affected by the Functional Groups.

Recently, quantum chemical calculations have been successfully applied to assess and predict the adsorption of pollutions by CNTs. Woods et al. [308] studied the adsorption of benzene derivatives by CNT using density functional theory and found that physical adsorption dominated through the interactions between the π orbitals of benzene derivatives and CNT. Tournus and Charlier [309] carried out an *ab initio* study of benzene adsorption on CNTs and found that the adsorption energies varied with the chiral angle of CNT. Zou et al. [310] systematically investigated the adsorption of cyclohexane, benzene derivatives, and polycyclic aromatic hydrocarbons on CNT using M05–2X of density-functional theory. The contribution of different functional groups on CNT to adsorption was quantified, and the calculated ΔG correlated well with the experiment values (Fig. A6).

With the development of molecular simulation methods, the adsorption of pollutants by CNTs can be predicted effectively using chemical calculations. With the number of aforementioned advantages: stronger chemical-nanotube interactions, tailored surface chemistry, rapid equilibrium rates, and high sorption capacity, CNTs were considered as superior sorbents for a wide range of organic chemicals and inorganic contaminants than the conventional ACs. However, for practical application in water treatment, the small particle size of CNTs will cause excessive pressure drops and the recovery of spent CNTs is a true challenge.

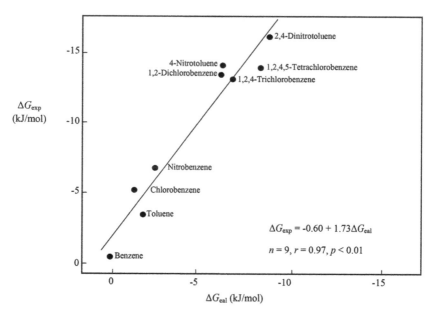

FIGURE A6 Regression of calculated ΔG_{exp} with experimental ΔG_{exp} values from nine aromatic compounds on SWCNT.

The macroscopic manipulation of CNT monolithic blocks via appropriate methods provide breakthrough for this bottleneck. Gui et al. [311] made a monolithic CNT sponge by chemical vapor deposition using ferrocene as precursor. The as-made CNT sponge had a randomly intertwined three-dimensional structure and displayed high porosity and very low density. The CNT sponge can float on oilcontaminated water and remove oil with a large adsorption capacity (80 to 180 times their own weight for a wide range of solvents and oils).The sponge had a tendency to move to the oil film area due to its high hydrophobicity, leading to the unique "floating-and-cleaning" capability that is very useful for spill cleanup.

In addition to serving as direct adsorbents, CNTs can also be used as excellent scaffold for macromolecules or metal oxides with intrinsic adsorption ability. The tunable surface chemistry and controllable pore size make CNTs good support for composite adsorbents. Examples of CNTs as scaffolds for pollutant removal include CNT decoration with iron oxide for europium adsorption [312], chitosan for methyl orange adsorption [313], polyaniline for malachite green adsorption [314], and ceria nanoparticles for chromium adsorption [315]. Maggini et al. [316] synthesized a supermolecular adsorbent by coating magnetic CNT with poly(vinylpyridine), whichexhibited excellent removal capacity of divalent metals from water, and the exhausted material can be separated using magnetic field and regenerated by acid treatment. The unique electrical properties of CNTs could be used for enhanced adsorption with electrochemical assist [317].

The mechanical flexibility and robustness, thermal stability and resistance to harsh environment endow CNTs with excellent application potential in water treatment. CNTs have the potential to serve as superior adsorbents for removal of both organic and inorganic contaminants from water systems. Nevertheless, there are several aspects that need to be evaluated prudently before the real application in water treatment facilities, including cost, reusability and the possibility of leakage into the environment.

A5 FILLED COMPOSITES

The mechanical behavior of carbon nanotubes is exciting since nanotubes are seen as the "ultimate" carbon fiber ever made. The traditional carbon fibers [318–319] have about 50 timesthe specific strength (strength/density) of steel and are excellent load-bearing reinforcements in composites. Nanotubes should then be ideal candidates for structural applications. Carbon fibers have been used as reinforcements in high strength, light weight, high performance composites; one can typically find these in a range of products ranging from expensive tennis rackets to spacecraft and aircraft body parts.

NASA has recently invested large amounts of money in developing carbon nanotube-based composites for applications such as the futuristic

Mars mission. Early theoretical work and recent experiments on individual nanotubes (mostly MWNTs) have confirmed that nanotubes are one of the stiffest structures ever made [12, 35, 43, 46, 320]. Since carbon–carbon covalent bonds are one of the strongest in nature, a structure based on a perfect arrangement of these bonds oriented along the axis of nanotubes would produce an exceedingly strong material. Many studies show that SWNTs could have a Young's modulus as high as 1Tpa [12], which is basically the in-plane value of defect free graphite. For MWNTs, the actual strength in practical situations would be further affected by the sliding of individual graphene cylinders with respect to each other. In fact, very recent experiments have evaluated the tensile strength of individual MWNTs using a nano-stressing stage located within a scanning electron microscope [40]. The nanotubes broke by a sword-in-sheath failure mode [319].

This failure mode corresponds to the sliding of the layers within the concentric MWNT assembly and the breaking of individual cylinders independently. The same failure modes have been observed previously in vapor grown carbon fibers [319].

The individual tubes in a rope could pull out by shearing along the rope axis, resulting in the final breakup of the rope, at stresses much below the tensile strength of individual nanotubes. Although testing of individual nanotubes is challenging, and requires specially designed stages and nanosize loading devices, some clever experiments have provided valuable insights into the mechanical behavior of nanotubes and have provided values for their modulus and strength. For example, in one of the earlier experiments, nanotubes projecting out onto holes in a TEM specimen grid were assumed to be equivalent to clamped homogeneous cantilevers; the horizontal vibrational amplitudes at the tube ends were measured from the luring of the images of the nanotube tips and were then related to the Young's modulus [43]. Recent experiments have also used atomic force microscopy to bend nanotubes attached to substrates and thus obtain quantitative information about their mechanical properties [35, 40, 44].

Most of the experiments done to date corroborate theoretical predictions suggesting the values of Young's modulus of nanotubes to be around 1 Tpa. Although the theoretical estimate for the tensile strength of individual SWNTs is about 300 GPa, the best experimental values (on MWNTs) are close to 50 GPa [40], which is still an order of magnitude higher than that of carbon fibers [318, 321].

The fracture and deformation behavior of nanotubes is intriguing. Simulations on SWNTs have suggested very interesting deformation behavior; highly deformed nanotubes were seen to switch reversibly into different morphological patterns with abrupt releases of energy. Nanotubes gets flattened, twisted and buckled as they deform. They sustain large strains (40%) in tension without showing signs of fracture [12].

The reversibility of deformations, such as buckling, has been recorded directly for MWNT, under TEM observations [322]. Flexibility of MWNTs depends on the number of layers that make up the nanotube walls; tubes with thinner walls tend to twist and flatten more easily. This flexibility is related to the in-plane flexibility of a planar graphene sheet and the ability for the carbon atoms to re-hybridize, with the degree of sp^2–sp^3 re-hybridization depending on the strain. Such flexibility of nanotubes under mechanical loading is important for their potential application as nanoprobes, for example, for use as tips of scanning probe microscopes.

Recently, an interesting mode of plastic behavior has been predicted in nanotubes [323]. It is suggested that pairs of 5–7(pentagon–heptagon) pair defects, called a Stone–Wales defect [324], in sp^2 carbon systems, are created at high strains in the nanotube lattice and that these defect pairs become mobile. This leads to a step-wise diameter reduction (localized necking) of the nanotube. These defect pairs become mobile. The separation of the defects creates local necking of the nanotube in the region where the defects have moved. In addition to localized necking, the region also changes lattice orientation (similar in effect to a dislocation passing through a crystal). This extraordinary behavior initiates necking but also introduces changes in helicity in the region where the defects have moved (similar to a change in lattice orientation when a dislocation passes through a crystal). This extraordinary behavior could lead to a unique nanotube application: a new type of probe, which responds to mechanical stress by changing its electronic character.

High temperature fracture of individual nanotubes under tensile loading, has been studied by molecular dynamics simulations [186]. Elastic stretching elongates the hexagons until, at high strain, some bonds are broken. This local defect is then redistributed over the entire surface, by bond saturation and surface reconstruction. The final result of this is that instead of fracturing, the nanotube lattice unravels into a linear chain of carbon atoms. Such behavior is extremely unusual in crystals and could play a role in increasing the toughness (by increasing the energy absorbed during deformation) of nanotube-filled ceramic composites during high temperature loading.

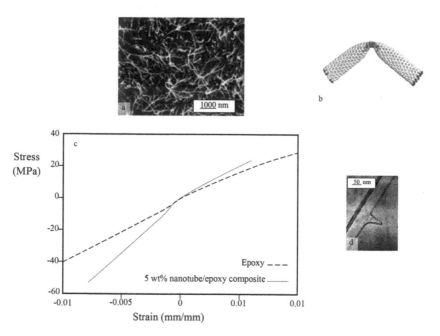

FIGURE A7   Results of the mechanical properties from MWNT-polymer (epoxy) composites. (a) SEM micrograph that shows good dispersion of MWNTs in the polymermatrix. The tubes are, however, elastically bent due to their highly flexible nature. Schematic of an elastically bent nanotube is shown in (b).Here, the strain is concentrated locally near the bend. (c) Stress–strain relationship observed during the tension/ compression testing of the nanotube-epoxy composite. It can be seen that the load transfer to the nanotube is higher during the compression cycle, because in tension the individual layers of the nanotubes slide with respect to each other. (d) TEM image of a thicker straight MWNT as well as a buckled MWNT in an epoxy matrix after loading.

The most important application of nanotubes based on their mechanical properties will be as reinforcements in composite materials. Although nanotube- filled polymer composites are an obvious materials application area, there have not been many successful experiments, which show the advantage of using nanotubes as fillers over traditional carbon fibers. The main problem is in creating a good interface between nanotubes and the polymer matrix and attaining good load transfer from the matrix to the nanotubes, during loading. The reason for this is essentially two-fold. First, nanotubes are atomically smooth and have nearly the same diameters and aspect ratios (length/diameter) as polymer chains. Second, nanotubes are almost always organized into aggregates,which behave differently in response to a load, as compared to individual nanotubes. There have been conflicting reports on the interface strength in nanotube-polymer composites [15, 70, 74, 76, 325].

Depending on the polymer used and processing conditions, the measured strength seems to vary. In some cases, fragmentation of the tubes has been observed, which an indication of a strong interface bonding is. In some cases, the effect of sliding of layers of MWNTs and easy pull-out are seen, suggesting poor interface bonding. Micro-Raman spectroscopy has validated the latter, suggesting that sliding of individual layers in MWNTs and *shearing* of individual tubes in SWNT ropes could be limiting factors for good load transfer, which is essential for making high strength composites.

To maximize the advantage of nanotubes as reinforcing structures in high strength composites, the aggregates needs to be broken up and dispersed or cross-linked to prevent slippage [74]. In addition, the surfaces of nanotubes have to be chemically modified (functionalized) to achieve strong interfaces between the surrounding polymer chains (Fig. A8).

There are certain advantages that have been realized in using carbon nanotubes for structural polymer (e.g., epoxy) composites. Nanotube reinforcements will increase the toughness of the composites by absorbing energy during their highly flexible elastic behavior. This will be especially important for nanotube-based ceramic matrix composites. An increase in fracture toughness on the order of 25% has been seen in nano-crystalline alumina nanotube (5% weight fraction) composites, without compromising on hardness [326]. Other interesting applications of nanotube-filled polymer films will be in adhesives where a decoration of nanotubes on the surface of the polymer films could alter the characteristics of the polymer chains due to interactions between the nanotubes and the polymer chains; the high surface area of the nanotube structures and their dimensions being nearly that of the linear dimensions of the polymer chains could give such nanocomposites new surface properties.

The low density of the nanotubes will clearly be an advantage for nanotube-based polymer composites, in comparison to short carbon fiber reinforced (random) composites. Nanotubes would also offer multifunctionality, such as increased electrical conduction. Nanotubes will also offer better performance during compressive loading in comparison to traditional carbon fibers due to their flexibility and low propensity for carbon nanotubes to fracture under compressive loads.

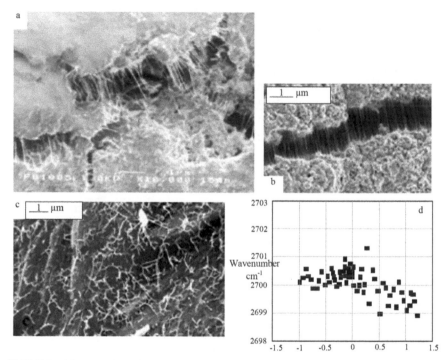

FIGURE A8 Results of mechanical properties measurements on SWNT-polymer composites. (a) SEM micrograph that shows a partially fractured surface of a SWNTepoxy composite, indicating stretched nanotubes extending across cracks. (b)A similar event illustrating the stretching and aligning of SWNT bundles across a long crack in SWNT-carbon soot composite. (c) SEM micrograph that shows the surface of a fractured SWNT-epoxy composite where the nanotube ropes have been completely pulled out and have fallen back on the fractured surface, forming a loose random network of interconnected ropes. (d)Results of micro-Raman spectroscopy that detects peak-shifts (in wave-numbers) as a function of strain. In both tension and compression of the SWNT-epoxy specimens, the peak shifts are negligible, suggesting no load transfer to the nanotubes during the loading of thecomposites.

Other than for structural composite applications, some of the unique properties of carbon nanotubes are being pursued by filling photo-active polymers with nanotubes. Recently, such a scheme has been demonstrated in a conjugated luminescent polymer, poly(m-phenylenevinylene-co-2,5-dioctoxyp-phenylenevinylene)(PPV), filled with MWNTs and SWNTs [327]. Nanotube/PPV composites have shown large increases in electrical conductivity (by nearly eight orders of magnitude)compared to the pristine polymer, with little loss in photoluminescence/electro-luminescence yield.

In addition, the composite is far more robust than the pure polymer regarding mechanical strength and photo-bleaching properties (breakdown of the polymer structure due to thermal effects). Preliminary studies indicate that the host polymer interacts weakly with the embedded nanotubes, but that the nanotubes act as nano-metric heat sinks, which prevent the build up of large local heating effects within the polymer matrix. While experimenting with the composites of conjugated polymers, such as PPV and nanotubes, a very interesting phenomenon has been recently observed [328]; it seems that the coiled morphology of the polymer chains helps to wrap around nanotubes suspended in dilute solutions of the polymer. This effect has been used to separate nanotubes from other carbonaceous material present in impure samples. Use of the non-linear optical and optical limiting properties of nanotubes has been reported for designing nanotube-polymer systems for optical applications, including photo-voltaic applications [329]. Functionalization of nanotubes and the doping of chemically modified nanotubes in low concentrations into photo-active polymers, such as PPV, have been shown to provide a means to alter the hole transport mechanism and hence the optical properties of the polymer.

Small loadings of nanotubes are used in these polymer systems to tune the color of emission when used in organic light emitting devices [330]. The interesting optical properties of nanotube-based composite systems arise from the low dimensionality and unique electronic band structure of nanotubes; such applications cannot be realized using larger micron-size carbon fibers (Fig. A9).

There are other less-explored areas where nanotube-polymer composites could be useful. For example, nanotube filled polymers could be useful in ElectroMagnetic Induction (EMI) shielding applications where carbon fibers have been used extensively [319]. Membranes for molecular separations (especially biomolecules) could be built from nanotube-polycarbonate systems, making use of the remarkable small pores sizes that exist in nanotubes. Very recently, work done at RPI suggests that composites made from nanotubes (MWNTs) and a biodegradable polymer (polylactic acid; PLA) act more efficiently than carbon fibers for osteointegration (growth of bone cells), especially under electrical stimulation of the composite. There are challenges to be overcome when processing nanotube composites. One of the biggest problems is dispersion.

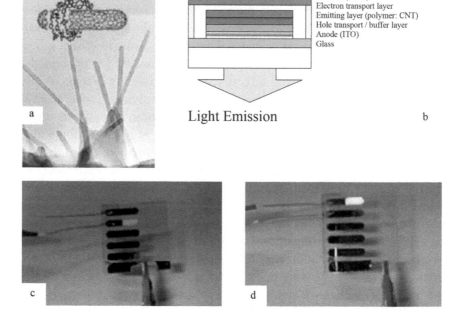

FIGURE A9   Results from the optical response of nanotube-doped polymers and their use in Organic Light Emitting Diodes (OLED). The construction of the OLED is shown in the schematic of *(top)*. The *bottom* figure shows emission from OLED structures. Nanotube doping tunes the emission color. With SWNTs in the buffer layer, holes are blocked and recombination takes place in the transport layer and the emission color is red [327].

It is extremely difficult to separate individual nanotubes during mixing with polymers or ceramic materials and this creates poor dispersion and clumping together of nanotubes, resulting in a drastic decrease in the strength of composites. By using high power ultrasound mixers and using surfactants with nanotubes during processing, good nanotube dispersion may be achieved, although the strengths of nanotube composites reported to date have not seen any drastic improvements over high modulus carbon fiber composites. Another problem is the difficulty in fabricating high weight fraction nanotube composites, considering the high surface area for nanotubes,which results in a very high viscosity for nanotube-polymer mixtures. Notwithstanding all these drawbacks, it needs to be said that the presence of nanotubes stiffens the matrix (the role is especially crucial at higher temperatures) and could be very useful as a matrix modifier [65], particularly for fabricating improved matrices useful for carbon fiber composites. The real role of nanotubes as an efficient reinforcing fiber will have to wait until we know how to manipulate the nanotube surfaces chemically to make strong interfaces between individual nanotubes (which are really the strongest material ever made) and the matrix materials. In the meanwhile, novel and unconventional uses of nanotubes will have to take the center stage.

A6   CARBON NANOTUBES IN HYDROGEN STORAGE

The area of hydrogen storage in carbon nanotubes remains active and controversial. Extraordinarily high and reversible hydrogen adsorption in SWNTcontaining materials [331–334] and graphite nanofibers (GNFs) [335] has been reported and has attracted considerable interest in both academia and industry. Table A1 summarizes the gravimetric hydrogen storage capacity reported by various groups [321]. However, many of these reports have not been independently verified. There is also a lack of understanding of the basic mechanism(s) of hydrogen storage in these materials.

Materials with high hydrogen storage capacities are desirable for energy storage applications. Metal hydrides and cryo-adsorption are the two commonly used means to store hydrogen, typically at high pressure and/ or low temperature. In metal hydrides, hydrogen is reversibly stored in the interstitial sites of the host lattice. The electrical energy is produced by direct electrochemical conversion. Hydrogen can also be stored in the gas phase in the metal hydrides. The relatively low gravimetric energy density has limited the application of metal hydride batteries. Because of their cylindrical and hollow geometry, and nanometer-scale diameters, it has been predicted that the carbon nanotubes can store liquid and gas in the inner cores through a capillary effect [284].

TABLE A1   Summary of Reported Garvimetric Storage of H_2 in Various Carbon Materials

Materials	Max. wt.% H_2	T (K)	P (Mpa)
SWNTs (low purity)	5–10	133	0.040
SWNTs (high purity)	4	300	0.040
CNFs (tubular)	11.26	298	11.35
CNFs(herringbone)	67.55	298	11.35
GNFs (platelet)	53.68	298	11.35
Graphite	4.52	298	11.35
GNFs	0.4	298	11.35
SWNTs (low purity)	8.25	80	7.18
SWNTs (50% pure)	4.2	300	10.1

A Temperature-Programmed Desorption (TPD) study on SWNT-containing material (0.1–0.2wt.% SWNT) estimates a gravimetric storage density of 5–10 wt.% SWNT when H2 exposures were carried out at 300 torr for 10 min at 277K followed by 3 min at 133K [336]. If all the hydrogen molecules are assumed to be inside the nanotubes, the reported density would imply a much higher packing density of H_2 inside the tubes than expected from the normal H_2–H_2 distance. The same group recently performed experiments on purified SWNTs and found essentially no H_2 absorption at 300K [333]. Upon cutting (opening) the nanotubes by an oxidation process, the amount of absorbed H_2 molecules increased to 4–5wt.%. A separate study on higher purity materials reports ~8wt.% of H_2 adsorption at 80 K, but using a much higher pressure of 100 atm. [337] suggesting that nanotubes have the highest hydrogen storage capacity of any carbon material. It is believed that hydrogen is first adsorbed on the outer surface of the crystalline ropes. An even higher hydrogen uptake, up to 14–20wt.%, at 20–400□C under ambient pressure was reported [336]in alkali-metal intercalated carbon nanotubes.

It is believed that in the intercalated systems, the alkali metal ions act as a catalytic center for H_2 dissociative adsorption. FTIR measurements show strong alkali–H and C–H stretching modes. An electrochemical absorption and desorption of hydrogen experiment performed on SWNT-containing materials (MER Co, containing a few percent of SWNTs) reported a capacity of 110 mAh/g at low discharge currents [331]. Measurements performed on relatively large amount materials (~50% purity, 500 mg) showed a hydrogen storage capacity of 4.2 wt.% when the samples were exposed to 10 MPa hydrogen at room temperature. About 80% of the absorbed H_2 could be released at room temperature [332].

The potential of achieving/exceeding the benchmark of 6.5 wt.% H_2 to system weight ratio set by the Department of Energy has generated considerable research activities in universities, major automobile companies and national laboratories. At this point it is still not clear whether carbon nanotubes will have real technological applications in the hydrogen storage applications area. The values reported in the literature will need to be verified on well-characterized materials under controlled conditions.

What is also lacking is a detailed understanding on the storage mechanism and the effect of materials processing on hydrogen storage. Perhaps the ongoing neutron scattering and proton nuclear magnetic resonance measurements will shed some light in this direction. In addition to hydrogen, carbon nanotubes readily absorb other gaseous species under ambient conditions, which often leads to drastic changes in their electronic properties [282, 338–339]. This environmental sensitivity is a double-edged sword. From the technological point of view, it can potentially be used for gas detection [282]. On the other hand, it makes very difficult to deduce the intrinsic properties of the nanotubes, as demonstrated by the recent transportand nuclear magnetic resonance [339] measurements. Care must be taken to remove the adsorbed species, which typically requires annealing the nanotubes at elevated temperatures under at least 6–10 torr dynamic vacuum.

METHODS FOR CARBON NANOTUBES SYNTHESIS

High temperature preparation techniques such as arc discharge or laser ablation were first used to produce CNTs but nowadays these methods have been replaced by low temperature chemical vapor deposition (CVD) techniques (<800°C), since the orientation, alignment, nanotube length, diameter, purity and density of CNTs can be precisely controlled in the latter [30, 340]. The most used methods and some of other non-standard techniques like liquid pyrolysis and bottom-up organic approach are discussed below.

Most of these methods require supporting gases and vacuum, but the growth at atmospheric pressure has been already reported [341]. However, gas-phase methods are volumetric and hence they are suitable for applications such as composite materials that require large quantities of nanotubes and industrial-scale synthesis methods to make them economically feasible. On the other hand, the disadvantages of gas-phase synthesis methods are low catalyst yields, where only a small percentage of catalysts form nanotubes, short catalyst lifetimes, and low catalyst number density[342]. Whatever CNT preparation method is applied, the CNTs are always produced with a number of impurities whose type and amount depend on the technique used. Most of above mentioned techniques produce powders which contain only a small fraction of CNTs and also other carbonaceous particles such as nanocrystalline graphite, amorphous carbon, fullerenes and different metals (typically Fe, Co, Mo or Ni) that were introduced as catalysts during the synthesis. These impurities interfere with most of the desired properties of CNTs and cause a serious impediment in detailed characterization and applications. Therefore, one of the most fundamental challenges in CNT science is the development of efficient and simple purification methods[343]. Some of the most commonCNT synthesis methods are discussed below:

B1   ARC DISCHARGE

Arc discharge belongs to the methods that use higher temperatures (above 1700°C) for CNT synthesis, which usually causes the growth of CNTs with fewer structural defects in comparison with other techniques. This method is discussed in following sections for different kind of carbon nanotubes.

B1.1   MWNTS SYNTHESIS

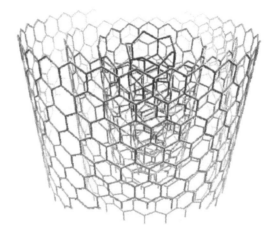

The arc discharge synthesis of MWNTs is very simple in the case when all growth conditions are ensured. The most used methods useDCarc discharge between two graphite usually water-cooled electrodes with diameters between 6 and 12 mm in a chamber filled with helium at sub-atmospheric pressure. Nevertheless, some other works with the use of hydrogen or methane atmosphere have been also reported. For example, Ebbesen and Ajayan use a variant of the standard arc-discharge technique also used by Iijima [1] for fullerene synthesis under He atmosphere to obtain first large-scale synthesis of CNTs. Under certain conditions, a pure nanotube and nanoscale particles in high yield were obtained. The purity and yield depended sensitively on the gas pressure in the reaction vessel [344]. Wang et al. [345] showed that different atmospheres markedly influence the final morphology of CNTs. They used DC arc discharge of graphite electrodes in He and methane. By evaporation under high-pressured CH_4 gas and high arc current, thick nanotubes embellished with many carbon nanoparticles were obtained. On the other hand, thin and long MWNTs were obtained under a CH_4 gas pressure of 50 Torr and an arc current of 20 A for the anode with a diameter of 6 mm. Moreover, Zhao et al. found that the variation of carbon nanotube morphology was more marked for the case of evaporation in CH4 gas than that in He gas [346]. In different work, Zhao et al. [347] used hydrogen gas atmosphere for preparation of fine and long MWNTs. By comparing with He and methane gases, a very big difference was found. Namely, little carbon smoke occurred in H_2 gas, but much more carbon smoke was observed for the evaporation in CH4 and He gases. Later they showed that evaporation of graphite electrodes in H_2 gas by DC arc discharge forms not only fine and long MWNTs but also graphene sheets deposited on the cathode [348].

Shimotani et al. reported synthesis of MWNTs using an arc discharge technique under He, ethanol, acetone and hexane atmosphere at various pressures (from 150 to 500 Torr). They concluded that arc discharges in the three organic atmospheres (ethanol, acetone and hexane) produce more MWNTs, by two times at least, than those in the He atmosphere. This can be explained as follows: contrary to helium, the acetone, ethanol and hexane can be ionized and the molecules can be decomposed into hydrogen and carbon atoms. These ionized species may contribute the synthesis of MWNTs, so the higher yield of CNTs is produced. They showed that in all the cases of organic molecular atmospheres, yields of MWNTs increase as the pressure increases up to 400 Torr [349].

Jiang et al. studied the influence of NH_3 atmosphere on the arc-discharge growth of CNTs and demonstrating that the arc-discharge method in NH_3 atmosphere is one highly efficient method for CNTs preparation. They concluded that there is no significant difference of the shapes and the structures between NH_3 atmosphere and other atmospheres such as He, H_2, etc. [350]. The consumption of anode during the process is faster than the growth of MWNTs layer on the cathode. Therefore the gap between the electrodes of 30 to 110 mm^2 surface area has to be held in the desired distance during the growth process (usually between 1 and 4 mm). This is ensured by one electrode constant feed that leads to a high yield and stable arc discharge growth process.

The arc discharge deposition is usually done as a DC arc discharge, but pulsed techniques were also reported. For example Parkansky et al. reported single-pulse arc production of near vertically oriented MWNTs deposited on the Ni/glass samples using a graphite counterelectrode in ambient air. MWNTs (typically 5–15 walls) with a diameter of about 10 nm and lengths of up to 3 mm were produced on the samples with a single 0.2 ms pulse [351]. Tsai et al. also used single-pulse discharge in air. They obtained MWNTs with the outer diameter of 17 nm and an inner diameter of 5 nm using a peak current of 2.5 A and a discharging time of 1000 ms[184]. Arc discharge is usually used for some non-standard CNTs deposition. Contrary to standard MWNTs deposition using a gas atmosphere there were reported several works involving arc discharge in liquid solutions. Jung et al. reported high yield synthesis of MWNTs by arc discharge in liquid nitrogen. They concluded that this technique can be a practical option for the large-scale synthesis of MWNTs with high purity [352]. A similar method was also used for MWNTs deposition by Sornsuwit and Maaithong[353]. Montoro et al. reported the synthesis of highquality SWNTs and MWNTs through arc-discharge in H_3VO_4 aqueous solution from pure graphite electrodes. DC arc discharge was generated between two high purity graphite electrodes. The high-resolution TEM images clearly showed that MWNTs are highly crystalline, with a well-ordered structure and free of defects. They obtain MWNTs with an outer diameter of 10–20 nm and an interlayer distance of approximately 0.35 nm between graphene layers [354].MWNTs were also synthesized in high yield by arc discharge in water between pure graphite electrodes by Guo et al. [355]. The production of carbon nanomaterials by arc discharge under water or liquid nitrogen was also reported by Xing et al. [356].

B1.2 SWNTS SYNTHESIS

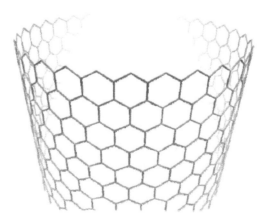

The arc discharge deposition of CNTs could be done without use of or with use of different catalyst precursors. Usually the MWNTs are produced when no catalyst is used. On the other hand, the SWNTs are produced when the transition metal catalyst is used. The process of SWNTs growth in arc discharge uses a composite anode, usually in hydrogen or argon atmosphere. The anode is made as a composition of graphite and a metal, such as Ni, Fe, Co, Pd, Ag, Pt, etc. or mixtures of Co, Fe, Ni with other elements like Co–Ni, Fe–Ni, Fe–No, Co–Cu, Ni–Cu, Ni–Ti, etc. The metal catalyst plays a significant role in the process yield. To ensure high efficiency, the process also needs to be held on a constant gap distance between the electrodes,which ensures stable current density and anode consumption rate. In this process, unwanted products such as MWNTs or fullerenes are usually produced too.

Firstly, the SWNTs growth process was described in two separate works by Iijima and Ichihashi [357] who presented SWNTs of 1 nm and Bethune et al. who described Co catalyzed growth of SWNTs. Bethune et al. reported that co-evaporation of carbon and cobalt in an arc generator leads to the formation of carbon nanotubes with very small diameters (about 1.2 nm) and walls made of a single atomic layer thick only[7]. Ajayan et al. also use Co catalyst for SWNTs synthesis of 1–2 nm diameter using arc discharge in He atmosphere [322]. One of the most used catalysts for SWNTs synthesis is nickel. Seraphin et al. studied the catalytic role of Ni, Pd, and Pt in the formation of carbon nanoclusters using DC arc discharge operated at 28 V, 70 A, and under a 550 Torr He atmosphere. They found out that nickel-filled anode stimulated the growth of SWNTs [358]. A similar method was used by Saito et al. who reported SWNTs growing radially from Ni fine particles [359]. Zhou et al. reported radially grown SWNTs synthesized using yttrium carbide loaded anode [353]. In 1996 Saito et al. reported the investigation of single-layered nanotubes produced with platinum-group metals (Ru, Rh, Pd, Os, Ir, Pt) using arc discharge. They reported that Rh, Pd, and Pt showed catalytic activity for growing SWNTs, but the other metals did not. Bundles of dense SWNTs with diameter 1.3–1.7 nm were extruding radially from metal particles for Rh and Pd; the sizes of core particles were 20–30 nm for Rh and 50–200 nm for Pd. In the case of Pt, one or few SWNTs (typically 1.3–2.0 nm in diameter and sometimes similar to 3 nm) grew from a tiny particle (of about 10 nm) [38]. In another work, Saito et al. reported SWNTs produced by the arc discharge method with Fe, Co, Ni, F/Ni, La, and Ce catalysts.

According to growth patterns and morphology of SWNTs, they divided the synthesis results into three groups: the tubes tangled with each other to form "highway junction" pattern for Co and Fe/Ni, long and thin tubes radially growing from Ni particles, and short and thick tubes growing from lanthanide compound particles[360].

The arc discharge method is still in use for SWNTs synthesis, but usually with a new approach. Chen et al. reported the FH (ferrum–hydrogen) arc discharge method. Using this method, SWNTs are produced by a hydrogen DC arc discharge with evaporation of carbon anode containing 1% Fe catalyst in H2–Ar mixture gas. The as-grown SWNTs have high crystallinity. An oxidation purification process of as-grown SWNTs with H_2O_2 has been developed to remove the coexisting Fe catalyst nanoparticles. As a result, SWNTs with purity higher than 90% have been achieved [361–362]. Zhao et al. looked for a cheap method for SWNTs synthesis. They successfully produced SWNTs in argon DC arc discharge from charcoal as carbon source and FeS (20 wt.%) as catalyst. According to SEM, TEM and Raman analysis, they achieved high purity SWNTs with diameter of about 1.2 nm. By this easy-to-get and relatively low cost material, the experimental results clearly indicated that charcoal has the opportunity of reducing the cost of SWNTs production [363].

In another work, Wang et al. studied the role of Mo on the growth of SWNTs in the arc discharge method. They incorporated Mo into Ni/Y–He and Fe–Ar/H_2, which are two typical arc systems. In both systems, Mo dramatically increased the yield of soot. The authors found that the purity of SWNTs did not change effectively for the Ni–Y/Mo–He system with the addition of Mo, but noticeable increment of purity was observed for Fe/Mo–Ar/H-2 system [364]. Li et al. [365] presented a possibility of SWNTs synthesis in air by pulsed arc discharge by preheating the catalyst to 600°C as an optimum that assists with the synthesis of SWNTs in air under pressure of 5–10 kPa. The SWNTs had a diameter of 1.5–2 nm and reached the length of several micrometers.

B1.3   DWNTS SYNTHESIS

The process of DWNTs deposition is more complicated than the production of SWNTs andMWNTs, but several successful attempts at methods for their preparation using arc discharge have been reported. Hutchison

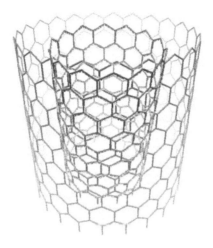

et al. first reported DWNTs an arc discharge technique in a mixture atmosphere of argon and hydrogen[366]. The anode was a graphite rod of 8.2 mm in diameter filled with catalyst. A mixture of Ni, Co, Fe and S was used as the catalyst. The obtained DWNTs formed into bundles as a rule. Occasionally, SWNTs were observed as a by-product. Sugai et al. reported new synthesis of high-quality DWNTs by the high-temperature pulsed arc discharge method using Y/Ni alloy catalysts [367]. Later, more sophisticated methods for DWNTs of higher quality appeared. DWNTs super bundles grown selectively above a bowl-like cathode by arc discharge in a hydrogen-free atmosphere were reported by Huang et al.[368] Their DWNTs can resist high-temperature (up to 720°C) oxidation in air without additional annealing even after acid treatment. This can be explained by an in situ defect-healing effect of the bowl-like cathode and the absence of reactive gases during arc discharge. Synthesis of DWNTs from coal in hydrogen-free atmosphere was also reported by Qiu et al.[369]. Qiu et al. [370] reported highly efficient and high scale synthesis of relatively perfect structural integrity DWNTs by an arc discharge method using trace halide (particularly potassium chloride) as a promoter in an iron sulfide catalyst. Both as synthesized DWNTs and purified DWNTs resisted to high temperature oxidation. It was proved that potassium chloride is a crucial factor for high yield formation of DWNTs with fewer defects.

Another work by Liu et al. reported preparation of DWNTs using nickel format dihydrate as an effective catalyst precursor for selectively synthesizing DWNTs with excellent oxidation resistance up to 800°C using a hydrogen arc discharge technique [371]. The synthesis of DWNTs from MWNTs by hydrogen arc discharge was reported by Li et al. DWNTs were synthesized in a large scale using graphite powders or MWNTs/carbon nanofibers as carbon source. They found that their DWNT product had higher purity than that from graphite powders. The results from HRTEM observations revealed that more than 80% of the CNTs were DWNTs and the rest were SWNTs. It was observed that the ends of the isolated DWNTs were uncapped and it was also found that cobalt as the dominant composition of the catalyst played a vital role in the growth of DWNTs by this method [372].

The general problem of produced CNTs is the presence of impurities that usually influence the final properties of CNTs as a material that could be used in some special application. Therefore, several papers reported on dealing with this problem. Acidic and thermal treatment, annealing, oxidation, filtration, ultrasonication and other techniques are used for CNTs purification. For example Ando et al. reported that easy purification of the MWNTs prepared by DC arc discharge of graphite electrodes in H_2 gas could be done by removal of coexisting carbon nanoparticles using infrared irradiation in a heating system in air at 500°C for 30 min [373]. The effect of calcination at different temperatures ranging from 300 to 600°C on MWNTs produced by DC arc discharge was studied by Pillai et al. They found that calcination in air at 400°C for 2 h is an efficient and simple method to eliminate carbonaceous impurities from the nanotube bundles with minimal damage to the tube walls and length[374]. The issue of CNT purification represents a wide field of investigation and reviews of CNTs purification have been already published, e.g. in ref [375].

B2   OTHER SYNTHESIS METHODS

B2.1   LASER ABLATION

The properties of CNTs prepared by the pulsed laser deposition process (PLD) are strongly dependent on many parameters such as: the laser properties (energy fluence, peak power, cw versus pulse, repetition rate and oscillation wavelength), the structural and chemical composition of the target material, the chamber pressure and the chemical composition, flow and pressure of the buffer gas, the substrate and ambient temperature and the distance between the target and the substrates. Laser ablation, as crucial step of PLD, is one of the superior methods to grow SWNTs with high-quality and high-purity.

In this method, which was first demonstrated by Smalley's group in 1995 [58], the principles and mechanisms are similar to the arc discharge with the difference that the energy is provided by a laser hitting a graphite pellet containing catalyst materials (usually nickel or cobalt)[376]. Almost all the lasers used for the ablation have been Nd: YAG and CO_2. For example, Zhang et al. prepared SWNTs by continuous wave CO_2 laser ablation without applying additional heat to the target. They observed that the average diameter of SWNTs produced by CO_2 laser increased with increasing laser power[377, 378]. Until now, the relationship between the excitation wavelength and the growth mechanisms of SWNTs has not been clarified. It may be expected that a UV laser creates a new species of nanoparticles and suggests a new generation mechanism of CNTs because the UV laser is superior in the photochemical ablation to the infrared laser,which is effective for photothermal ablation.

Lebel et al. [379] synthesized SWNTs using the UV-laser (KrF excimer) ablation of a graphite target appropriately doped with Co/Ni metal catalyst. In their work, they tested as-prepared SWNTs as a reinforcing agent of polyurethane. Kusaba and Tsunawaki used XeCl excimer laser with the oscillation wavelength of 308 nm to irradiate a graphite containing Co and Ni at various temperatures and they found that laser ablation at 1623 K produced the highest yield of SWNTs with the diameter between 1.2 and 1.7 nm and the length of 2 mm or above[380]. Recently, Stramel et al. have successfully applied commercial MWNTS and MWNTs–polystyrene targets (PSNTs) for deposition of composite thin films onto silicon substrates using PLD with a pulsed, diode pumped, Tm:Ho:LuLF laser (a laser host material LuLF ($LuLiF_4$) is doped with holmium and thulium in order to reach a laser light production in the vicinity of 2 mm)[381].

They found that usage of pure MWNTs targets gives rise to a thin film containing much higher quality MWNTs compared to PSNTs targets. Similarly, Bonaccorso et al. [382] prepared MWNTs thin films deposited by PLD techniques (with Nd: YAG laser) ablating commercially polystyrene-nanotubes pellets on alumina substrates.

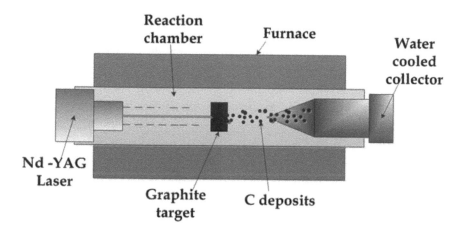

FIGURE B1   Schematic representation of Laser ablation method used for carbon nanotube synthesis.

B2.2   CHEMICAL VAPOR DEPOSITION

Catalytic chemical vapor deposition (CCVD)(either thermal[383] or plasma enhanced (PE)) is now the standard method for the CNTs production. Moreover, there are trends to use other CVD techniques, like water assisted CVD,[384, 385] oxygen assisted CVD [386], hot-filament (HFCVD) [387], microwave plasma (MPECVD)[388, 389] or radiofrequency CVD (RF-CVD) [390]. CCVD is considered to be an economically viable process for large scale and quite pure CNTs production compared with laser ablation. The main advantages of CVD are easy control of the reaction course and high purity of the obtained material, etc.[391].

Recently, Fotopoulos and Xanthakis discussed the traditionally accepted models, which are base growth and tip growth. In addition, they mentioned a hypothesis that SWNTs are produced by base growth only, that is, the cap is formed first and then by a lift off process the CNT is created by addition of carbon atoms at the base. They refer to recent in situ video rate TEM studies,which have revealed that the base growth of SWNT in thermal CVD is accompanied by a considerable deformation of the Ni catalyst nanoparticle and the creation of a subsurface carbon layer. These effects may be produced by the adsorption on the catalyst nanoparticle during pyrolysis[392]. In order to produce SWNTs, the size of the nanoparticle catalyst must be smaller than about 3 nm. The function of the catalyst in the CVD process is the decomposition of carbon source via either plasma irradiation (plasma-enhanced CVD, PECVD) or heat (thermal CVD) and its new nucleation to form CNTs.

The most frequently used catalysts are transition metals, primarily Fe, Co, or Ni [79]. Sometimes, the traditionally used catalysts are further doped with other metals (e.g., with Au) [393]. Concerning the carbon source, the most preferred in CVD are hydrocarbons such as methane[394], ethane[395], ethylene[396], acetylene [397], xylene[353], eventually their mixture [87],isobutene [88], or ethanol [89,90]. In the case of gaseous carbon source, the CNTs growth efficiency strongly depends on the reactivity and concentration of gas phase intermediates produced together with reactive species and free radicals as a result of hydrocarbon decomposition. Thus, it can be expected that the most efficient intermediates, which have the potential of chemisorption or physisorption on the catalyst surface to initiate CNT growth, should be produced in the gas phase[398].

Commonly used substrates are Ni, Si, SiO_2, Cu, Cu/Ti/Si, stainless steel or glass, rarely $CaCO_3$; graphite and tungsten foil or other substrates were also tested [399]. A special type of substrate, mesoporous silica, was also tested since it might play a templating role in guiding the initial nanotube growth. For example, Zhu et al. [400] reported a CCVD synthesis of DWNTs over supported metal catalysts decomposed from Fe and Co onmesoporous silica.

They obtained bundles of tubes with a relatively high percentage of DWNTs in areas where tubular layered structures could be clearly resolved. Moreover, the crystal-like alignment of very uniform DWNTs was observed.Similarly, Ramesh et al. [401] succeeded in high-yield selective CVD synthesis of DWNTs over Fe/Co loaded high-temperature stable mesoporous silica. Another substrate, zeolites, was studied by Hiraoka et al. [402]. They used CCVD of acetylene over well-dispersed metal particles (typically Co/Fe binary system) embedded in heat-resistant zeolites at temperatures above 900°C for selective synthesis of DWNTs.

The choice of catalyst is one of the most important parameters affecting the CNTs growth. Therefore, its preparation is also a crucial step in CNTs synthesis. The influence of the composition and the morphology of the catalyst nanoparticles on CNTs growth by CVD are summarized in Ref. [402]. Flahaut et al. [403] reported the influence of catalyst preparation conditions for the synthesis of CNTs by CCVD. In their work, the catalysts were prepared by the combustion route using either urea or citric acid as the fuel. They found that the milder combustion conditions obtained in the case of citric acid can either limit the formation of carbon nanofibers or increase the selectivity of the CCVD synthesis towards CNTs with fewer walls, depending on the catalyst composition.

B2.3   PLASMA ENHANCED CHEMICAL VAPOR DEPOSITION (PECVD)

Plasma enhanced chemical vapor deposition (PECVD) is a suitable method for synthesis of CNTs hybrid materials and modification of their surface properties. Lim et al. reviewed the application of PECVD in the production and modification of CNTs. They emphasize the usage of the PECVD method for SWNTs growing at low temperatures and make an effort to better understand plasma chemistry and modeling[349]. PECVD can be also used in several different modes: radio frequency (RF-PECVD), direct current (DC-PECVD), diffusion (DPECVD) or microwave (MWPECVD). Kim and Gangloff [404] demonstrated the low-temperature (480–612°C) synthesis of CNTs on different metallic under layers(i.e., NiV, Ir, Ag, Pt, W, and Ta) using DPECVD. They used a Fe/Al bilayer as the catalyst.

Wang and Moore prepared vertically aligned CNTs using FeNi or Fe sputtered catalyst layers on glass substrates by RF- or DC-PECVD. They compared the CNTs growth mechanisms using both methods with respect to gas flow rate, plasma power and catalysts. They explained why RF-PECVD provided more efficient decomposition of gas molecules than DC-PECVD by plasma theory. The major difference between RF- and DCPECVD was the higher concentration of reactive radicals in the former. However, in DC-PECVD, the CNT growth was well aligned vertically. They found that FeNi thin film catalysts exhibited higher activity and better wetting ability than the Fe island thin film catalysts[405].

Like in thermal CVD, numerous catalyst types to improve the yield and the quality of CNTs production are also applied in PECVD. For example, Luais et al. [406] prepared spherical Ni nanoparticles film Ni $(NO_3)_2$ as a starting material, which was used as a catalyst. The diameter of Ni nanoparticles was about 50 nm. This catalyst was further used for synthesis of vertically aligned CNTs by PECVD in an electron cyclotron resonance chamber using a gas mixture of C_2H_2/NH_3 at 520°C. The average thickness of the CNTs film was about 1 mm and the CNTs diameter was around 50 nm. After CNTs preparation, their surface was functionalized with oxygenated and aminated groups using microwave plasma to make them suitable for future biosensing applications. Moreover, they found that the plasma treatment was a very effective way to retain the CNTs aligned forest structure of electrode surface.

Haffner et al. [407] demonstrated the fabrication of a biocompatible system of CNTs electrodes by PECVD using ferritin as the catalyst material. Ferritin consists of a small Fe_2O_3 compound core with a diameter in the nanometer range, enclosed by a protein shell a few nanometers thick. Treated with oxygen plasma, amino acids around the ferritin cores were removed, the iron cores were automatically separated from each other and dense vertically aligned CNTs grew from the well-separated iron cores. For possible application of these CNTs electrodes in neuroimplants, which is based on flexible temperature sensitive substrates (like artificial mica), it is important to reach low temperatures during the preparation process (down to 450°C).

Pd was also tested as a catalyst material in the work of Vollebregt et al. [408], who prepared vertically self-aligned CNTs and CNFs. The authors compared two preparation methods with various conditions and catalysts (Pd, Ni, Fe, Co) as follows: PECVD at 450°C to 500°C and atmospheric-pressure chemical vapor deposition (APCVD) between 450°C and 640°C. High-density self-aligned CNTs were obtained using APCVD and Pd as the catalyst, while Co and Fe resulted in random growth. TEM revealed that the CNTs grown by Pd with PECVD formed large bundles of tubes, while Ni formed large-diameter CNFs. The authors found that the CNTs grown using Pd or Ni were of low quality compared with those grown by Co and Fe.

However, some works were published with no usage of catalysts for CNTs growing. Qu et al. [408] prepared new hybrid material consisting of spontaneous assembly of carbon nanospheres on aligned or non-aligned SWNTs using the PECVD method. The carbon nanospheres were formed with a uniform size of 30–60 nm. The formation of these spheres is a catalyst-free process and strongly depends on the applied plasma power and other factors. This heterojunction structure based on different types of carbon seems to be promising as a building complex system for various applications.

Analogous to various catalysts, different substrates for CNTs preparation can be also used in the PECVD process. For example, Duy et al. [409]

reported the fabrication of CNTs on Ni-coated stainless steel or Si substrates using DC-PECVD. The synthesized CNTs have a diameter of about 30 nm and a length of about 1.2 mm. They found that CNTs grown on the stainless steel substrates were more uniform compared with those grown on the Si substrates. Moreover, they showed the potential of CNTs in field emission applications, especially CNT-based cold-cathode X-ray tubes.

B2.4   CVD METHODS FOR UNIFORM VERTICALLY ALIGNED CNTS SYNTHESIS

The growth of vertically aligned carbon nanotube forests is studied extensively because it represents one of the highest yield methods of nanotube growth[410]. The aim of several works is to fabricate vertically aligned CNTs with homogeneous distribution on the surfaces and high uniformity. It is evident to use nanolithography to create pattern of the catalyst on the surface on which CNTs are formed using CVD methods. In the work of Kim et al., the Si wafer was used as a substrate and Ni as a catalyst deposited on a diffusion barrier from Ni/Ti. Ni dots catalyst of 1.6 mm and about 200 nm was patterned using UV and e-beam lithography, respectively. The method of CNTsformation uses the triode PECVD reactor with gas ratio $C_2H_2/(H_2$ or $NH_3)$ at 620°C. The positive ions in the cathode sheath of the plasma can force the CNTs to grow perpendicular to the substrate. The diameter of created CNTs depended on Ni dots size[411].

The most frequently used method of aligned CNTs formation (Fig. B3) is the usage of nanoporous anodized aluminum oxide (AAO) as template for CNTs growth. In the work of Kim et al. [412], the CNTs were grown on AAO/Si substrate. A catalytic metal layer was formed on the Si wafer by direct deposition. Two types (A and B) of nanoporous aluminum templates were used for the study of the growth characteristics of CNTs. Type A was aluminum of 500 nm thickness, which was deposited on a silicon wafer. Type B was aluminum on Co-coated silicon wafer. The thickness of the Colayer was 100°C. The pore diameter and pore depth was approximately 33 nm and 210 nm for Type A and 5 nm and 220 nm for Type B, respectively. After the pore widening process, the pore was enlarged to approximately 60 nm and 33 nm for Type A and Type B, respectively. The CNTs growth was carried out on the AAO at temperature below 550°C by DC-PECVD. An acetylene gas was used as a carbon source and an ammonia gas was used as a dilution and catalytic gas. The DC plasma was applied to grow vertically aligned CNTs. The CNT growth with PECVD on Type A was quite different from the CNT growth with thermal CVD because CNTs were not grown on the AAO/Si barrier.

Graphitization of the CNTs was very poor compared to the CNTs grown on glass substrate using thermal CVD. The CNTs grown on the barrier between pores do not look like nanotubes but carbon nanofibers. In the case of Type B the length of CNT is almost the same as the pore depth because CNTs did not grow on the AAO/Si barrier, but only on the catalyst on the bottom of the pores. CNTs grew on AAO/Si without a catalyst, while there was no overgrowth of CNTs on AAO/Si with a catalyst. In synthesis of CNTs using AAO template with/without a Co layer, both alumina and Co can work as a catalyst with flowing acetylene.

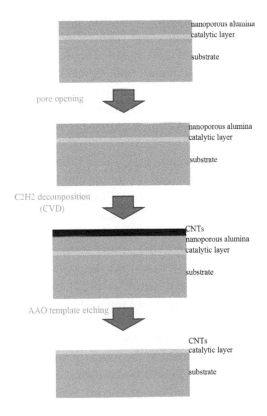

FIGURE B3   Process of CNTs growth using the AAO template.

B2.5   LIQUID PYROLYSIS

The aerosol pyrolysis process is a catalytic CVD-based method involving pyrolysis of mixed liquid aerosols composed of both liquid hydrocarbon and catalyst precursor. Byeon et al. [413] developed a new aerosol-assisted chemical vapor deposition (AACVD) process to synthesize vertically aligned CNTs arrays with outstanding height (4.38 mm) with very low metal contents in a short time (20 min) without supporting materials and water-assistance. An essential part of this technique was in situ formation of metal catalyst nanoparticles via pyrolysis of ferrocene–ethanol aerosol right before CNTs synthesis.

Jeong et al. [414] presented an ultrasonic evaporator atomizing the mixed liquid solution for MWNTs production by the thermal pyrolysis process. They produced aligned and clean CNTs, which can be easily controlled in a cost-effective manner. A similar approach for nitrogen-doped CNTs with the tunable structure and high yield production by ultrasonic spray pyrolysis was done by Liu et al.[415]. Khatri et al. [416] reported SWNTs synthesis using ethanol and bimetallic catalyst of cobalt and molybdenum acetates by an ultrasonic spray pyrolysis method on silicon substrates at 850°C [140]. Later, this author focused his research on zeolites powder as catalyst supporting material for SWNTs production using ultrasonic spray pyrolysis.

In another study, Camarena et al. [417] prepared MWNTs by spray pyrolysis using toluene as the carbon source and ferrocene as the catalyst. Sadeghian [418] reported preparation of MWNTs by spray pyrolysis, using hexane as a carbon source and ferrocene as a catalyst precursor. Clean and aligned MWNTs produced by aerosol pyrolysis of mixed liquid aerosols composed of both liquid hydrocarbon (toluene or cyclohexane) and catalyst precursor (ferrocene) were also reported by Pinault et al.[419]. In another work, Nebol'sin and Vorob'ev [420] studied CNTs growth via catalytic pyrolysis of acetylene. They found that surface free energy plays a key role in determining the catalytic activity of the liquid droplet on the CNT tip and is responsible for the constant nanotube diameter.

An interesting study describing the usage of a less common liquid carbon source, namely various pinene components isolated from turpentine, for MWNTs production by spray pyrolysis was recently published by Lara-Romero et al.[421]. Next, a green natural carbon source for CNTs fabrication, Neem oil extracted from the seeds of the Neem-Azadirachta indica, was tested in the work of Kumar et al.[422]. Similarly, coconut oil can be also used as a natural renewable precursor for MWNTs synthesis[423]. Quan et al. reported a new and interesting way of CNTs synthesis through waste pyrolysis oil. This process is based on treatment of waste pyrolysis liquid from printed circuit board (PCB) waste, which contains high concentrations of phenol and phenol derivatives. Hence, it can be applied as a carbon source in the preparation of advanced carbonaceous materials like CNTs. First, the pyrolysis oil is prepared by pyrolysis of PCB waste at 600°C. In the second step, the product is polymerized in a formaldehyde solution to synthesize pyrolysis oil-based resin,which is used as the precursor CNTs. Finally, this resin was mixed with ferrocene and homogenized in ethanol. After alcohol evaporation, the mixture was ground into fine powder, loaded on a ceramic boat and placed inside a stainless steel tubular reactor. The mixture was heated to 200°C in air with 1 h soaking time, and then up to 900°C in a flow of N_2 with holding periods for 1 h at 900°C. The resulting CNTs had hollow cores with outer diameter of 338 nm and wall thickness of 86 nm and most of them were filled with metal nanoparticles or nanorods. X-Ray diffraction revealed that CNTs had an amorphous structure[424].

B2.6   SOLID STATE PYROLYSIS

Nowadays, solid-state pyrolysis for CNTs synthesis is less frequently used compared to previously mention ones. Kucukayan et al.[425] synthesized MWNTs through pyrolysis of the sulfuric acid-carbonized byproduct of sucrose. They observed the presence of sulfur in catalyst particles trapped inside nanotubes, but no sulfur was present in the side-walls of the CNTs. Clauss et al. [426] thermally decomposed two nitrogen-rich iron salts, ferric ferrocyanide (Prussian Blue, "PB") and iron melonate ("FeM") in a microwave oven, which was used to heat a molybdenum wire after being coated with the precursor and protected from ambient atmosphere. While the PB-precursor did not give any nanotube containing products, the FeM-precursor furnished tubular carbon nanostructures in a reproducible manner.

This result may be due to the graphite-like nature of the melonate anions presented in FeM[426]. Kuang et al. [427] synthesized straight CNTs in large scale through thermal CVD by pyrolysis of two mixed metal phthalocyanines with a certain amount of S at 800–950°C. The as-synthesized CNTs were 15–35 nm in diameter and 200–800 nm in length, quite straight and well-graphitized with nearly no defects. Two kinds of mixed transition metal phthalocyanines (M (II)Pc, M ¼ Fe, Co) were used as the carbon source as well as catalysts favoring the growth of the straight CNTs. Du et al. [428] prepared MWNTs through the solid-phase transformation of metal-containing glass-like carbon nanoparticles by heating at temperatures of 800–1000°C. From microscopic observations on the morphologies and structures of the nanotubes and various intermediate objects, it is shown that the transformation occurs by nanoparticles first assembling into wire-like nanostructures, and then transforming into nanotubes via particle-particle coalescence and structural crystallization.

B2.7 FLAME PYROLYSIS

This technique is presented very uniquely by the research group of Liu et al. as a new method for mass CNTs production using simple equipment and experimental conditions. The authors called it V-type pyrolysis flame. They captured successfully CNTs with less impurities and high yield using carbon monoxide as the carbon source. Acetylene/air premixed gas provided heat by combustion. Pentacarbonyl was used as the catalyst and hydrogen/helium premixed gas acted as diluted and protection gas. The diameter of obtained CNTs was approximately between 10 nm and 20 nm, and its length was dozens of microns. Moreover they studied the effect of sampling time, hydrogen and helium to the CNTs growth process [429, 430].

B3 BOTTOM-UP ORGANIC APPROACH

The bottom-up approach to integrate vertically PECVD grown MWNTs into multilevel interconnects in silicon integratedcircuit manufacturing from patterned catalyst spot was reported first by Li et al.[431]. More recently, Jasti and Bertozzi [432] described in their frontier article the potential advantages, recent advances, and challenges that lie ahead for the bottom-up organic synthesis of homogeneous CNTs with well-defined structures.

Current synthetic methods used for CNTs fabrication produce mixtures of structures with varying physical properties. Jasti and Bertozzi [432] demonstrated the CNTs synthesis with control of chirality, which relies on using hoop-shaped carbon macrocycles, that is, small fragments of CNTs that retain information regarding chirality and diameter, as templates for CNTs synthesis. Their strategy lies in two basic areas: the synthesis of aromatic macrocyclic templates and the development of polymerization reactions to extend these templates into longer CNTs.

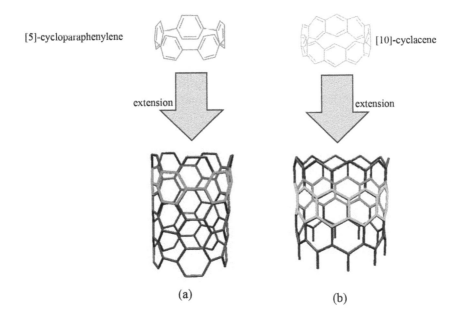

FIGURE B4 Bottom-up, organic synthesis approach to CNTs with discrete chirality.

This approach is particularly attractive because it can be used for synthesis of both zigzag and armchair CNTs of different diameters, as well as of chiral CNTs with various helical pitches. For example, a (5, 5) armchair CNT can be constructed by fusing additional phenyl rings to cycloparaphenylene (Fig. B4.a). In similar fashion, a (10, 0) zigzag CNT can be constructed from cyclacene (Fig. B4.b).

Since the publication work of Jasti and Bertozzi, several approaches for carbon macrocycles as a step toward the bottom-up synthesis of CNTs with selective chirality have been reported. Omachi et al. reported a modular and size-selective synthesis of [14]-, [15]-, and [16]-cycloparaphenylenes for selective synthesis of [n,n] type SWNTs [433].Concise synthesis of [12] cycloparaphenylene and its crystal structure was presented by Segawa et al.[434]. A simple and realistic model for the shortest sidewall segments of chiral SWNTs has been designed, and one of the chiral carbon nanorings, cyclo[13]paraphenylene-2,6-naphthylene for chemical synthesis of chiral CNTs has been successfully synthesized by Omachi et al.[435].

Selective and random synthesis of a [8]–[13]-cycloparaphenylenes mixture was prepared in good combined yields by mixing biphenyl and terphenyl precursors with platinum sources by Iwamoto et al.[436]. Finally Fort and Scott also presented interesting groundwork for the selective solvent-free growth of uniform diameter armchair CNTs by gas-phase Diels–Alder cyclo-addition of benzyne to an aromatic hydrocarbon bay region on the rims of suitable cylindrical hydrocarbon templates followed by rearomatizations and thermal cyclodehydrogenations to join adjacent benzo groups[437].

REFERENCES

1. Iijima, S. (1991). Helical Microtubules of Graphitic Carbon, Nature, *354(6348)*, 56–58.
2. Dresselhaus, M., Dresselhaus, G., & Eklund, P. (1996). Science of Fullerenes and Carbon Nanotubes Their Properties and Applications, Academic Press.
3. Saito, R., Dresselhaus, G., & Dresselhaus, M. (1998). Physical Properties of Carbon Nanotubes, *4*, World Scientific, 517.
4. Harris, P., & Harris P. (2001) Carbon Nanotubes and Related Structures, New Materials for the Twenty-First Century, Cambridge University Press, 279.
5. Wagner, H., & Vaia, R. (2004). Nanocomposites Issues at the Interface, Materials Today, *7(11)*, 38–42.
6. Zeng, Q., Yu, A., & Lu, G. (2008). Multiscale Modeling and Simulation of Polymer Nanocomposites, Progress in Polymer Science, *33(2)*, 191–269.
7. Bethune, D. et al. (1993). Cobalt-Catalyzed Growth of Carbon Nanotubes with Single-Atomic-Layer Walls.
8. Dresselhaus, M., Dresselhaus, G., & Saito, R. (1995). Physics of Carbon Nanotubes, Carbon, *33(7)*, 883–891.
9. Thostenson, E., Ren, Z., & Chou, T. (2001). Advances in the Science and Technology of Carbon Nanotubes and their Composites, a Review Composites Science and Technology, *61(13)*, 1899–1912.
10. Yakobson, B., & Avouris, P. (2001). Mechanical Properties of Carbon Nanotubes, in Carbon Nanotubes, Springer, 287–327.
11. Dresselhaus, M., & Avouris, P. (2001). Introduction to Carbon Materials Research, in Carbon Nanotubes, Springer, 90.
12. Yakobson, B., Brabec, C., & Bernholc, J. (1996). Nanomechanics of Carbon Tubes Instabilities Beyond Linear Response Physical Review Letters, *76(14)*, 2511–2514.
13. Ajayan, P., Schadler, L., & Braun, P. (2006). Nanocomposite Science and Technology, John Wiley & Sons, 550.
14. Li, J., et al. (2007). Correlations Between Percolation Threshold, Dispersion State, and Aspect Ratio of Carbon Nanotubes, Advanced Functional Materials, *17(16)*, 3207–3215.
15. Ajayan, P. et al. (1994). Aligned Carbon Nanotube Arrays Formed by Cutting a Polymer Resin-Nanotube Composite, Science, *265(5176)*, 1212–1214.
16. Du, J., Bai, J., & Cheng, H. (2007). The Present Status and Key Problems of Carbon Nanotube Based Polymer Composites, Express Polymer Letters, *1(5)*, 253–273.
17. Brown, T. (2009). Chemistry the Central Science, Pearson Education, 543.
18. Yu, M., Yakobson, B., & Ruoff, R. (2000). Controlled Sliding and Pullout of Nested Shells in Individual Multi-walled Carbon Nanotubes, The Journal of Physical Chemistry B, *104(37)*, 8764–8767.
19. Fennimore, A. et al. (2003). Rotational Actuators Based on Carbon Nanotubes, Nature, *424(6947)*, 408–410.

20. Saito, R., et al. (1992). Electronic Structure of Chiral Graphene Tubules, Applied Physics Letters, *60(18)*, 2204–2206.
21. Fujita, M., et al. (1992). Formation of General Fullerenes by their Projection on a Honeycomb Lattice, Physical Review B, *45(23)*, 13834.
22. Dresselhaus, M., Dresselhaus, G., & Eklund, P. (1993). Fullerenes Journal of Materials Research, *8(08)*, 2054–2097.
23. Yuklyosi, K. (1977). Encyclopedic Dictionary of Mathematics, MIT Press, Cambridge.
24. Iijima, S. (1993). Growth of Carbon Nanotubes, Materials Science and Engineering B, *19(1)*, 172–180.
25. Dravid, V., et al. (1993). Buckytubes and Derivatives their Growth and Implications for Buckyball Formation, Science, *259(5101)*, 1601–1604.
26. Ichihashi, T., & Ando, Y. (1992). Pentagons, Heptagons and Negative Curvature in Graphite Microtubule Growth, Nature, *356(6372)*, 776–778.
27. Saito, Y. et al. (1993). Interlayer Spacings in Carbon Nanotubes, Physical Review B, *48*, 1907–1909.
28. Zhou, O. et al. (1994). Defects in Carbon Nanostructures, Science, *263(5154)*, 1744–1747.
29. Kiang, C. et al. (1998). Size Effects in Carbon Nanotubes, Physical Review Letters, *81(9)*, 1869.
30. Amelinckx, S. et al. (1995). A Structure Model and Growth Mechanism for Multishell Carbon Nanotubes, Science, *267(5202)*, 1334–1338.
31. Gerard Lavin, J. et al. (2002). Scrolls and Nested Tubes in Multi-wall Carbon Nanotubes, Carbon, *40(7)*, 1123–1130.
32. Popov, V., Van Doren, V., & Balkanski, M. (2000). Elastic Properties of Crystals of Single-Walled Carbon Nanotubes, Solid State Communications, *114(7)*, 395–399.
33. Hernandez, E. et al. (1998). Elastic Properties of C and B x C y N z Composite Nanotubes, Physical Review Letters, *80(20)*, 4502.
34. Treacy, M., Ebbesen, T., & Gibson, J. (1996). Exceptionally High Young's Modulus Observed for Individual Carbon Nanotubes, 76–83.
35. Wong, E., Sheehan, P., & Lieber, C. (1997). Nanobeam Mechanics Elasticity, Strength, and Toughness of Nanorods and Nanotubes, Science, *277(5334)*, 1971–1975.
36. Salvetat, J., et al. (1999). Elastic and Shear Moduli of Single-Walled Carbon Nanotube Ropes, Physical Review Letters, *82(5)*, 944.
37. Walters, D. et al. (1999). Elastic Strain of Freely Suspended Single-Wall Carbon Nanotube Ropes, Applied Physics Letters, *74(25)*, 3803–3805.
38. Qian, D., et al. (2002). Mechanics of Carbon Nanotubes, Applied Mechanics Reviews, *55(6)*, 495–533.
39. Yu, M. et al. (2000). Tensile Loading of Ropes of Single Wall Carbon Nanotubes and Their Mechanical Properties, Physical Review Letters, *84(24)*, 5552.
40. Yu, M. et al. (2000). Strength and Breaking Mechanism of Multi-walled Carbon Nanotubes Under Tensile Load, Science, *287(5453)*, 637–640.
41. Xie, S. et al. (2000). Mechanical and Physical Properties on Carbon Nanotube, Journal of Physics and Chemistry of Solids, *61(7)*, 1153–1158.

42. Wong, S. et al. (1998). Carbon Nanotube Tips High-Resolution Probes for Imaging Biological Systems. Journal of the American Chemical Society, *120(3)*, 603–604.

43. Treacy, M., Ebbesen, T., & Gibson, J. (1996). Exceptionally High Young's Modulus Observed for Individual Carbon Nanotubes.

44. Falvo, M. et al. (1997). Bending and Buckling of Carbon Nanotubes Under Large Strain, Nature, *389(6651)*, 582–584.

45. Bower, C. et al. (1999). Deformation of Carbon Nanotubes in Nanotube-Polymer Composites, Applied Physics Letters, *74(22)*, 3317–3319.

46. Overney, G., Zhong, W., & Tomanek, D. (1993). Structural Rigidity and Low Frequency Vibrational Modes of Long Carbon Tubules, Zeitschrift Für Physik D Atoms, Molecules and Clusters, *27(1)*, 93–96.

47. Lu, J. (1997). Elastic Properties of Single and Multilayered Nanotubes, Journal of Physics and Chemistry of Solids, *58(11)*, 1649–1652.

48. Yakobson, B. et al. (1997). High Strain Rate Fracture and C-chain Unraveling in Carbon Nanotubes, Computational Materials Science, *8(4)*, 341–348.

49. Bernholc, J. et al. (1998). Theory of Growth and Mechanical Properties of Nanotubes, Applied Physics A Materials Science & Processing, *67(1)*, 39–46.

50. Iijima, S. et al. (1996). Structural Flexibility of Carbon Nanotubes, The Journal of Chemical Physics, *104(5)*, 2089–2092.

51. Ru, C. (2000). Effective Bending Stiffness of Carbon Nanotubes, Physical Review B, *62(15)*, 9973.

52. Vaccarini, L. et al. (2000). Mechanical and Electronic Properties of Carbon and Boron-Nitride Nanotubes, Carbon, *38(11)*, 1681–1690.

53. Al-Jishi, R., & Dresselhaus, G. (1982). Lattice-Dynamical Model for Graphite, Physical Review B, *26*, 4514–4522.

54. Yakobson, B., Samsonidze, G., & Samsonidze, G. (2000). Atomistic Theory of Mechanical Relaxation in Fullerene Nanotubes, Carbon, *38(11)*, 1675–1680.

55. Journet, C. et al. (1997). Large-Scale Production Of Single-Walled Carbon Nanotubes by the Electric-Arc Technique, Nature, *388(6644)*, 756–758.

56. Thess, A. et al. (1996). Crystalline Ropes of Metallic Carbon Nanotubes, Science-AAAS-Weekly Paper Edition, *273(5274)*, 483–487.

57. Ru, C. (2000). Elastic Buckling of Single-Walled Carbon Nanotube Ropes Under High Pressure, Physical Review B, *62(15)*, 10405.

58. Popov, V., Van Doren, V., & Balkanski, M. (1999). Lattice Dynamics of Single-Walled Carbon Nanotubes, Physical Review B, *59(13)*, 8355.

59. Ruoff, R., & Lorents, D. (1995). Mechanical and Thermal Properties of Carbon Nanotubes, Carbon, *33(7)*, 925–930.

60. Govindjee, S., & Sackman, J. (1999). On the Use of Continuum Mechanics to Estimate the Properties of Nanotubes, Solid State Communications, *110(4)*, 227–230.

61. Ru, C. (2000). Effect of van der Waals Forces on Axial Buckling of a Double-Walled Carbon Nanotube, Journal of Applied Physics, *87(10)*, 7227–7231.

62. Ru, C. (2001). Axially Compressed Buckling of a Doublewalled Carbon Nanotube Embedded in an Elastic Medium, Journal of the Mechanics and Physics of Solids, *49(6)*, 1265–1279.

63. Ru, C. (2001). Degraded Axial Buckling Strain of Multi-walled Carbon Nanotubes Due to Interlayer Slips, Journal of Applied Physics, *89(6)*, 3426–3433.

64. Kolmogorov, A., & Crespi, V. (2000). Smoothest Bearings Interlayer Sliding in Multi-walled Carbon Nanotubes, Physical Review Letters, *85(22)*, 4727–4730.

65. Shaffer, M., & Windle, A. (1999). Fabrication and Characterization of Carbon Nanotube/Poly (Vinyl Alcohol) Composites, Advanced Materials, *11(11)*, 937–941.

66. Qian, D. et al. (2000). Load Transfer and Deformation Mechanisms in Carbon Nanotube-Polystyrene Composites, Applied Physics Letters, *76(20)*, 2868–2870.

67. Jia, Z. et al. (1999). Study on Poly (Methyl Methacrylate)/Carbon Nanotube Composites, Materials Science and Engineering A, *271(1)*, 395–400.

68. Gong, X. et al. (2000). Surfactant-Assisted Processing of Carbon Nanotube/Polymer Composites, Chemistry of Materials, *12(4)*, 1049–1052.

69. Lordi, V., & Yao, N. (2000). Molecular Mechanics of Binding in Carbon-Nanotube-Polymer Composites, Journal of Materials Research, *15(12)*, 2770–2779.

70. Wagner, H. et al. (1998). Stress-Induced Fragmentation of Multi-wall Carbon Nanotubes in a Polymer Matrix, Applied Physics Letters, *72(2)*, 188–190.

71. Lourie, O., & Wagner, H. (1998). Transmission Electron Microscopy Observations of Fracture of Single-Wall Carbon Nanotubes Under Axial Tension, Applied Physics Letters, *73(24)*, 3527–3529.

72. Lourie, O., Cox, D., & Wagner, H. (1998). Buckling and Collapse of Embedded Carbon Nanotubes, Physical Review Letters, *81*, 1638–1641.

73. Cooper, C., Young, R., & Halsall, M. (2001). Investigation into the Deformation of Carbon Nanotubes and Their Composites Through the Use of Raman spectroscopy, Composites Part A Applied Science and Manufacturing, *32(3)*, 401–411.

74. Ajayan, P. et al. (2000). Single-Walled Carbon Nanotube–Polymer Composites Strength and Weakness, Advanced Materials, *12(10)*, 750–753.

75. Schadler, L., Giannaris, S., & Ajayan, P. (1998). Load Transfer in Carbon Nanotube Epoxy Composites, Applied Physics Letters, *73(26)*, 3842–3844.

76. Jin, L., Bower, C., & Zhou, O. (1998). Alignment of Carbon Nanotubes in a Polymer Matrix by Mechanical Stretching, Applied Physics Letters, *73(9)*, 1197–1199.

77. Haggenmueller, R. et al. (2000). Aligned Single-Wall Carbon Nanotubes in Composites by Melt Processing Methods, Chemical Physics Letters, *330(3)*, 219–225.

78. Andrews, R. et al. (1999). Nanotube Composite Carbon Fibers, Applied Physics Letters, *75(9)*, 1329–1331.

79. Gommans, H. et al. (2000). Fibers of Aligned Single-Walled Carbon Nanotubes, Polarized Raman Spectroscopy, Journal of Applied Physics, *88(5)*, 2509–2514.

80. Vigolo, B. et al. (2000). Macroscopic Fibers and Ribbons of Oriented Carbon Nanotubes, Science, *290(5495)*, 1331–1334.

81. Ru, C. (2000). Column Buckling of Multi-walled Carbon Nanotubes with Interlayer Radial Displacements, Physical Review B, *62(24)*, 16962.

82. Wang, C. et al. (2006). Buckling of Double-Walled Carbon Nanotubes Modeled by Solid Shell Elements, Journal of Applied Physics, *99(11)*, 114317.

83. Han, Q., & Lu, G. (2003). Torsional Buckling of a Double-Walled Carbon Nanotube Embedded in an Elastic Medium, European Journal of Mechanics-A/Solids, *22(6)*, 875–883.

84. Zhou, W. et al. (2006). Copper Catalyzing Growth of Single-Walled Carbon Nanotubes on Substrates, Nano Letters, *6(12)*, 2987–2990.

85. Han, S., Liu, X., & Zhou, C. (2005). Template-Free Directional Growth of Single-Walled Carbon Nanotubes on a-and r-Plane Sapphire, Journal of the American Chemical Society, *127(15)*, 5294–5295.

86. Wang, X., Yang, H., & Yin, X. (2005). Axially Critical Load of Multi-wall Carbon Nanotubes Under Thermal Environment, Journal of Thermal Stresses, *28(2)*, 185–196.

87. Leung, A. et al. (2006). Postbuckling of Carbon Nanotubes by Atomic-Scale Finite Element, Journal of Applied Physics, *99(12)*, 124308.

88. Yao, X., & Han, Q. (2008). Torsional Buckling and Postbuckling Equilibrium Path of Double-Walled Carbon Nanotubes, Composites Science and Technology, *68(1)*, 113–120.

89. Muc, A. (2011). Modelling of Carbon Nanotubes Behaviour with the Use of a Thin Shell Theory, Journal of Theoretical and Applied Mechanics, *49*, 531–540.

90. Muc, A. (2010). Design and Identification Methods of Effective Mechanical Properties for Carbon Nanotubes, Materials & Design, *31(4)*, 1671–1675.

91. Muc, A., Banas, A., & lgorzata, M. (2013). Chwa, Free Vibrations of Carbon Nanotubes with Defects, Mechanics and Mechanical Engineering, *17(2)*, 157–166.

92. Liew, K., & Wang, Q. (2007). Analysis of Wave Propagation in Carbon Nanotubes Via Elastic Shell Theories, International Journal of Engineering Science, *45(2)*, 227–241.

93. Hu, Y, et al. (2008). Nonlocal Shell Model for Elastic Wave Propagation in Single-and Double-Walled Carbon Nanotubes, Journal of the Mechanics and Physics of Solids, *56(12)*, 3475–3485.

94. Natsuki, T., Ni, Q., & Endo, M. (2008). Analysis of the Vibration Characteristics of Double-Walled Carbon Nanotubes, Carbon, *46(12)*, 1570–1573.

95. Ghorbanpourarani, A. et al. (2010). Transverse Vibration of Short Carbon Nanotubes Using Cylindrical Shell and Beam Models, Proceedings of the Institution of Mechanical Engineers, Part C Journal of Mechanical Engineering Science, *224(3)*, 745–756.

96. Tylikowski, A. (2008). Instability of Thermally Induced Vibrations of Carbon Nanotubes, Archive of Applied Mechanics, *78(1)*, 49–60.

97. Wang, J. et al. (2007). Capacitance Properties of Single Wall Carbon Nanotube/Polypyrrole Composite Films, Composites Science and Technology, *67(14)*, 2981–2985.

98. Lee, et al. (1998). Simulation of Polymer Melt Intercalation in Layered Nanocomposites, The Journal of Chemical Physics, *109(23)*, 10321–10330.

99. Smith, G. et al. (2002). A Molecular Dynamics Simulation Study of the Viscoelastic Properties of Polymer Nanocomposites, The Journal of Chemical Physics, *117(20)*, 9478–9489.

100. Zeng, Q. et al. (2003). Molecular Dynamics Simulation of Organic-Inorganic Nanocomposites, Layering Behavior and Interlayer Structure of Organoclays, Chemistry of Materials, *15(25)*, 4732–4738.
101. Zeng, Q. et al. Interfacial Structure and Interactions in Clay-Polymer Nanocomposites, 18–33.
102. Vacatello, M. (2003). Predicting the Molecular Arrangements in Polymer□Based Nanocomposites. Macromolecular Theory and Simulations, *12(1)*, 86–91.
103. Allen, M., & Tildesley, D. (1987). Computer Simulation of Liquids (1989) New York Oxford, *385*, 88–98.
104. Frenkel, D., & Smit, B. (2001). Understanding Molecular Simulation, from Algorithms to Applications, Academic press, 430.
105. Metropolis, N. et al. (2004). Equation of State Calculations by Fast Computing Machines, The Journal of Chemical Physics, *21(6)*, 1087–1092.
106. Hoogerbrugge, P., & Koelman, J. (1992). Simulating Microscopic Hydrodynamic Phenomena with Dissipative Particle Dynamics, EPL (Europhysics Letters), *19(3)*, 155.
107. Gibson, J., Chen, K., & Chynoweth, S. (1998). Simulation of Particle Adsorption onto a Polymer-Coated Surface Using the Dissipative Particle Dynamics Method, Journal of Colloid and Interface Science, *206(2)*, 464–474.
108. Dzwinel, W., & Yuen, D. (2000). A Two-Level, Discrete Particle Approach for Large-Scale Simulation of Colloidal Aggregates, International Journal of Modern Physics C, *11(05)*, 1037–1061.
109. Dzwinel, W., Yuen, D., & Boryczko, K. (2002). Mesoscopic Dynamics of Colloids Simulated with Dissipative Particle Dynamics and Fluid Particle Model, Molecular Modeling Annual, *8(1)*, 33–43.
110. Chen, S., & Doolen, G. (1998). Lattice Boltzmann Method for Fluid Flows, Annual Review of Fluid Mechanics, *30(1)*, 329–364.
111. Cahn, J. (1961). On Spinodal Decomposition, Acta Metallurgica, *9(9)*, 795–801.
112. Cahn, J. (2004). Free Energy of a Nonuniform System II Thermodynamic Basis, The Journal of Chemical Physics, *30(5)*, 1121–1124.
113. Lee, B., Douglas, J., & Glotzer, S. (1999). Filler-Induced Composition Waves in Phase-Separating Polymer Blends, Physical Review E, *60(5)*, 5812–5830.
114. Ginzburg, V. et al. (1999). Simulation of Hard Particles in a Phase-Separating Binary Mixture arXiv Preprint Cond-Mat/9905284, 18–28.
115. Qiu, F. et al. (1999). Phase Separation Under Shear of Binary Mixtures Containing Hard Particles, Langmuir, *15(15)*, 4952–4956.
116. He, G., & Balazs, A. (2005). Modeling the Dynamic Behavior of Mixtures of Diblock Copolymers and Dipolar Nanoparticles, Journal of Computational and Theoretical Nanoscience, *2(1)*, 99–107.
117. Altevogt, P. et al. (1999). The MesoDyn Project Software for Mesoscale Chemical Engineering, Journal of Molecular Structure THEOCHEM, *463(1)*, 139–143.
118. Morita, H., Kawakatsu, T., & Doi, M. (2001). Dynamic Density Functional Study on the Structure of Thin Polymer Blend Films with a Free Surface, Macromolecules, *34(25)*, 8777–8783.
119. Powell IV, A., & Arroyave, R. (2008). Open Source Software for Materials and Process Modeling, JOM, *60(5)*, 32–39.

120. Ginzburg, V. et al. (2000). Modeling the Dynamic Behavior of Diblock Copolymer/ Particle Composites, Macromolecules, *33(16)*, 6140–6147.

121. Tandon, G., & Weng, G. (1984). The Effect of Aspect Ratio of Inclusions on the Elastic Properties of Unidirectionally Aligned Composites, Polymer Composites, *5(4)*, 327–333.

122. Odegard, G., Pipes, R., & Hubert, P. (2004). Comparison of Two Models of SWCN Polymer Composites, Composites Science and Technology, *64(7)*, 1011–1020.

123. Odegard, G. et al. (2003). Constitutive Modeling of Nanotube-Reinforced Polymer Composite, Composites Science and Technology, *63(11)*, 1671–1687.

124. Pipes, R., & Hubert, P. (2002). Helical Carbon Nanotube Arrays, Mechanical Properties, Composites Science and Technology, *62(3)*, 419–428.

125. Rudd, R., & Broughton, J. (2000). Concurrent Coupling of Length Scales in Solid State Systems, Physica Status Solidi (b), *217(1)*, 251–291.

126. Starr, F., & Glotzer, S. (2000). Simulations of Filled Polymers on Multiple Length Scales in MRS Proceedings Cambridge Univ Press.

127. Glotzer, S., & Starr, F. (2001). Towards Multiscale Simulations of Filled and Nano-filled Polymers, in AIChE Symposium Series, New York, American Institute of Chemical Engineers (1998).

128. Mori, T., & Tanaka, K. (1973). Average Stress in Matrix and Average Elastic Energy of Materials with Misfitting Inclusions, Acta Metallurgica, *21(5)*, 571–574.

129. Fisher, F., Bradshaw, R., & Brinson, L. (2003). Fiber Waviness in Nanotube-Reinforced Polymer Composites-I Modulus Predictions Using Effective Nanotube Properties, Composites Science and Technology, *63(11)*, 1689–1703.

130. Bradshaw, R., Fisher, F., & Brinson, L. (2003). Fiber Waviness in Nanotube-Reinforced Polymer Composites-II Modeling Via Numerical Approximation of the Dilute Strain Concentration Tensor, Composites Science and Technology, *63(11)*, 1705–1722.

131. Liu, Y., & Chen, X. (2003). Evaluations of the Effective Material Properties of Carbon Nanotube-Based Composites Using a Nanoscale Representative Volume Element, Mechanics of Materials, *35(1)*, 69–81.

132. Liu, Y., & Chen, X. (2002). Modeling and Analysis of Carbon Nanotube-Based Composites Using the FEM and BEM, Submitted to CMES Computer Modeling in Engineering and Science, 88–97.

133. Liu, Y., Luo, J., & Xu, N. (2000). Modeling of Interphases in Fiber-Reinforced Composites Under Transverse Loading Using the Boundary Element Method, Journal of Applied Mechanics, *67(1)*, 41–49.

134. Fu, Y. et al. (1998). A Fast Solution Method for Three-Dimensional Many-Particle Problems of Linear Elasticity, International Journal for Numerical Methods in Engineering, *42(7)*, 1215–1229.

135. Nishimura, N., Yoshida, K., & Kobayashi, S. (1999). A Fast Multipole Boundary Integral Equation Method for Crack Problems in 3D, Engineering Analysis with Boundary Elements, *23(1)*, 97–105.

136. Qian, D., Liu, W., & Ruoff, R. (2001). Mechanics of C60 in Nanotubes, The Journal of Physical Chemistry B, *105(44)*, 10753–10758.

137. Chen, X., & Liu, Y. (2004). Square Representative Volume Elements for Evaluating the Effective Material Properties of Carbon Nanotube-Based Composites, Computational Materials Science, *29(1)*, 1–11.

138. Wan, H., Delale, F., & Shen, L. (2005). Effect of CNT Length and CNT-Matrix Interphase in Carbon Nanotube (CNT) Reinforced Composites, Mechanics Research Communications, *32(5)*, 481–489.

139. Shi, D. et al. (2005). Multiscale Analysis of Fracture of Carbon Nanotubes Embedded in Composites, International Journal of Fracture, *134(3–4)*, 369–386.

140. Buryachenko, V. et al. (2005). Multi-Scale Mechanics of Nanocomposites Including Interface Experimental and Numerical Investigation, Composites Science and Technology, *65(15)*, 2435–2465.

141. Li, C., & Chou, T. (2003). Multiscale Modeling of Carbon Nanotube Reinforced Polymer Composites, Journal of Nanoscience and Nanotechnology, *3(5)*, 423–430.

142. Li, C., & Chou, T. (2006). Multiscale Modeling of Compressive Behavior of Carbon Nanotube/Polymer Composites, Composites Science and Technology, *66(14)*, 2409–2414.

143. Anumandla, V., & Gibson, R. (2006). A Comprehensive Closed Form Micromechanics Model for Estimating the Elastic Modulus of Nanotube-Reinforced Composites, Composites Part A Applied Science and Manufacturing, *37(12)*, 2178–2185.

144. Seidel, G., & Lagoudas, D. (2006). Micromechanical Analysis of the Effective Elastic Properties of Carbon Nanotube Reinforced Composites, Mechanics of Materials, *38(8)*, 884–907.

145. Hashin, Z., & Rosen, B. (1964). The Elastic Moduli of Fiber-Reinforced Materials, Journal of Applied Mechanics, *31(2)*, 223–232.

146. Christensen, R., & Lo, K. (1979). Solutions for Effective Shear Properties in Three Phase Sphere and Cylinder Models, Journal of the Mechanics and Physics of Solids, *27(4)*, 315–330.

147. Selmi, A. et al. (2007). Prediction of the Elastic Properties of Single Walled Carbon Nanotube Reinforced Polymers a Comparative Study of Several Micromechanical Models, Composites Science and Technology, *67(10)*, 2071–2084.

148. Luo, D., Wang, W., & Takao, Y. (2007). Effects of the Distribution and Geometry of Carbon Nanotubes on the Macroscopic Stiffness and Microscopic Stresses of Nanocomposites, Composites Science and Technology, *67(14)*, 2947–2958.

149. Fu, S. et al. (2000). On the Elastic Stress Transfer and Longitudinal Modulus of Unidirectional Multi-Short-Fiber Composites, Composites Science and Technology, *60(16)*, 3001–3012.

150. Tserpes, K. et al. (2008). Multi-Scale Modeling of Tensile Behavior of Carbon Nanotube-Reinforced Composites, Theoretical and Applied Fracture Mechanics, *49(1)*, 51–60.

151. Spanos, P., & Kontsos, A. (2008). A Multiscale Monte Carlo Finite Element Method for Determining Mechanical Properties of Polymer Nanocomposites, Probabilistic Engineering Mechanics, *23(4)*, 456–470.

152. Zhu, J. et al. (2004). Reinforcing Epoxy Polymer Composites Through Covalent Integration of Functionalized Nanotubes, Advanced Functional Materials, *14(7)*, 643–648.

153. Paiva, M. et al. (2004). Mechanical and Morphological Characterization of Polymer–Carbon Nanocomposites from Functionalized Carbon Nanotubes, Carbon,. *42(14),* 2849–2854.

154. Bhuiyan, M., et al. (2013). Defining the Lower and Upper Limit of the Effective Modulus of CNT/Polypropylene Composites Through Integration of Modeling and Experiments, Composite Structures, *95,* 80–87.

155. Papanikos, P., Nikolopoulos, D., & Tserpes, K. (2008). Equivalent Beams for Carbon Nanotubes, Computational Materials Science, *43(2),* 345–352.

156. Affdl, J., & Kardos, J. (1976). The Halpin□Tsai Equations a Review Polymer Engineering & Science, *16(5),* 344–352.

157. Landel, R., & Nielsen, L. (1993). Mechanical Properties of Polymers and Composites, CRC Press, 553.

158. Tucker, III C., & Liang, E. (1999). Stiffness Predictions for Unidirectional Short-Fiber Composites, Review and Evaluation, Composites Science and Technology, *59(5),* 655–671.

159. Shokrich, M., & Rafiee, R. (2010). Investigation of Nanotube Length Effect on the Reinforcement Efficiency in Carbon Nanotube Based Composites, Composite Structures, *92(10),* 2415–2420.

160. Shokrieh, M., & Rafiee, R. (2010). On the Tensile Behavior of an Embedded Carbon Nanotube in Polymer Matrix with Non-Bonded Interphase Region, Composite Structures, *92(3),* 647–652.

161. Haghi, A., & Zaikov, G. (2014). Carbon Nanotubes and Related Structures, in Handbook of Research on Functional Materials Principles, Capabilities and Limitations, CRC Press, 147–159.

162. Esawi, A., & Farag, M. (2007). Carbon Nanotube Reinforced Composites, Potential and Current Challenges, Materials & Design, *28(9),* 2394–2401.

163. Chowdhury, S. et al. (2012). Modeling the Effect of Statistical Variations in Length and Diameter of Randomly Oriented CNTs on the Properties of CNT Reinforced Nanocomposites, Composites Part B Engineering, *43(4),* 1756–1762.

164. Gou, J. et al. (2004). Computational and Experimental Study of Interfacial Bonding of Single-Walled Nanotube Reinforced Composites Computational Materials Science, *31(3),* 225–236.

165. Li, S. et al. (2005). Electrical Properties of Soluble Carbon Nanotube/Polymer Composite Films, Chemistry of Materials, *17(1),* 130–135.

166. Liu, Y., & Chen, X. (2007). Continuum Models of Carbon Nanotube-Based Composites Using the Boundary Element Method, Electronic Journal of Boundary Elements, *1(2),* 18–33.

167. Thostenson, E., & Chou, T. (2002). Aligned Multi-walled Carbon Nanotube-Reinforced Composites Processing and Mechanical Characterization, Journal of Physics D Applied Physics, *35(16),* 77–90.

168. Daniel Wagner, H. (2002). Nanotube-Polymer Adhesion A Mechanics Approach Chemical Physics Letters, *361(1),* 57–61.

169. Valentini, L. et al. (2003). Physical and Mechanical Behavior of Single□Walled Carbon Nanotube/Polypropylene/Ethylene–Propylene–Diene Rubber Nanocomposites, Journal of Applied Polymer Science, *89(10),* 2657–2663.

170. Paipetis, A. et al. (1999). Stress Transfer from the Matrix to the Fibre in a Fragmentation Test Raman Experiments and Analytical Modeling, Journal of Composite Materials, *33(4)*, 377–399.

171. Liao, K., & Li, S. (2001). Interfacial Characteristics of a Carbon Nanotube-Polystyrene Composite System, Applied Physics Letters, *79(25)*, 4225–4227.

172. Andrews, R., & Weisenberger, M. (2004). Carbon Nanotube Polymer Composites, Current Opinion in Solid State and Materials Science, *8(1)*, 31–37.

173. Frankland, S., & Harik, V. (2003). Analysis of Carbon Nanotube Pull-Out From a Polymer Matrix, Surface Science, *525(1)*, L103-L108.

174. Wernik, J., Cornwell-Mott, B., & Meguid, S. (2012). Determination of the Interfacial Properties of Carbon Nanotube Reinforced Polymer Composites Using Atomistic-Based Continuum Model, International Journal of Solids and Structures, *49(13)*, 1852–1863.

175. Wernik, J., & Meguid, S. (2011). Multiscale Modeling of the Nonlinear Response of Nano-Reinforced Polymers, Acta Mechanica, *217(1–2)*, 1–16.

176. Yang, S. et al. (2012). Multiscale Modeling of Size-Dependent Elastic Properties of Carbon Nanotube/Polymer Nanocomposites with Interfacial Imperfections, Polymer, *53(2)*, 623–633.

177. Ayatollahi, M., Shadlou, S., & Shokrieh, M. (2011). Multiscale Modeling for Mechanical Properties of Carbon Nanotube Reinforced Nanocomposites Subjected to Different Types of Loading, Composite Structures, *93(9)*, 2250–2259.

178. Jiang, L. et al. (2006). A Cohesive Law for Carbon Nanotube/Polymer Interfaces Based on the van der Waals Force, Journal of the Mechanics and Physics of Solids, *54(11)*, 2436–2452.

179. Tan, H. et al. (2007). The Effect of van der Waals-Based Interface Cohesive Law on Carbon Nanotube-Reinforced Composite Materials, Composites Science and Technology, *67(14)*, 2941–2946.

180. Zalamea, L., Kim, H., & Pipes, R. (2007). Stress Transfer in Multi-walled Carbon Nanotubes, Composites Science and Technology, *67(15)*, 3425–3433.

181. Shen, G., Namilae, S., & Chandra, N. (2006). Load Transfer Issues in the Tensile and Compressive Behavior of Multi-wall Carbon Nanotubes, Materials Science and Engineering A, *429(1)*, 66–73.

182. Gao, X., & Li, K. (2005). A Shear-Lag Model for Carbon Nanotube-Reinforced Polymer Composites, International Journal of Solids and Structures, *42(5)*, 1649–1667.

183. Li, C., & Chou, T A. (2003). Structural Mechanics Approach for the Analysis of Carbon Nanotubes, International Journal of Solids and Structures, *40(10)*, 2487–2499.

184. Tsai, Y. et al. (2009). Production of Carbon Nanotubes by Single-Pulse Discharge in Air, Journal of Materials Processing Technology, *209(9)*, 4413–4416.

185. Belytschko, T. et al. (2002). Atomistic Simulations of Nanotube Fracture, Physical Review B, *65(23)*, 235430.

186. Yakobson, B., Brabec, C., & Bernholc, J. (1996). Structural Mechanics of Carbon Nanotubes, From Continuum Elasticity to Atomistic Fracture, Journal of Computer-Aided Materials Design, 3(1–3), 173–182.

187. Huang, Y., Wu, J., & Hwang, K. (2006). Thickness of Graphene and Single-Wall Carbon Nanotubes, Physical Review B, *74(24)*, 245413.

188. Wu, J., Hwang, K., & Huang, Y. (2008). An Atomistic-Based Finite-Deformation Shell Theory for Single-Wall Carbon Nanotubes, Journal of the Mechanics and Physics of Solids, *56(1)*, 279–292.

189. Peng, J. et al. (2008). Can a Single-Wall Carbon Nanotube be Modeled as a Thin Shell? Journal of the Mechanics and Physics of Solids, *56(6)*, 2213–2224.

190. Kalamkarov, A. et al. (2006). Analytical and Numerical Techniques to Predict Carbon Nanotubes Properties, International Journal of Solids and Structures, *43(22)*, 6832–6854.

191. Li, Y. et al. (2005). Adsorption Thermodynamic, Kinetic and Desorption Studies of Pb^{2+} on Carbon Nanotubes, Water Research, *39(4)*, 605–609.

192. Kalamkarov, A. (1987). On the Determination of Effective Characteristics of Cellular Plates and Shells of Periodic Structure, Mechanics of Solids.

193. Kalamkarov, A. (1992). Composite and Reinforced Elements of Construction, Wiley Chichester, 340.

194. Kalamkarov, A., & Kolpakov, A. (1997). Analysis, Design, and Optimization of Composite Structures J Wiley & Sons.

195. Kalamkarov, A., & Georgiades, A. (2004). Asymptotic Homogenization Models for Smart Composite Plates with Rapidly Varying Thickness, Part I–Theory International Journal for Multiscale Computational Engineering, *2(1)*.

196. Reddy, J. (2003). Mechanics of Laminated Composite Plates and Shells Theory and Analysis, CRC Press, 430.

197. Kalamkarov, A., Veedu, V., & GhasemiNejhad, M. (2005). Mechanical Properties Modeling of Carbon Single-Walled Nanotubes, An Asymptotic Homogenization Method, Journal of Computational and Theoretical Nanoscience, *2(1)*, 124–131.

198. Machida, T. et al. (2003). Coherent Control of Nuclear-Spin System in a Quantum-Hall Device, Applied Physics Letters, *82(3)*, 409–411.

199. Hart, J., & Rappe, A. (1992). van der Waals Functional Forms for Molecular Simulations, The Journal of Chemical Physics, *97(2)*, 1109–1115.

200. Tersoff, J., & Ruoff, R. (1994). Structural Properties of a Carbon-Nanotube Crystal, Physical Review Letters, *73(5)*, 676.

201. Brenner, D. (1990). Empirical Potential for Hydrocarbons for Use in Simulating the Chemical Vapor Deposition of Diamond Films, Physical Review B, *42(15)*, 9458.

202. Cornell, W. et al. (1995). A Second Generation Force Field for the Simulation of Proteins, Nucleic Acids, and Organic Molecules, Journal of the American Chemical Society, *117(19)*, 5179–5197.

203. Gelin, B. (1994). Molecular Modeling of Polymer Structures and Properties Hanser/Gardner Publications.

204. Walther, M., Fischer, B., & Uhd Jepsen, P. (2003). Noncovalent Intermolecular Forces in Polycrystalline and Amorphous Saccharides in the Far Infrared Chemical Physics, *288(2)*, 261–268.

205. László, I., & Rassat, A. (2003). The Geometric Structure of Deformed Nanotubes and the Topological Coordinates Journal of Chemical Information and Computer Sciences, *43(2)*, 519–524.

206. Jorgensen, W., & Severance, D. (1991). Chemical Chameleons Hydrogen Bonding with Imides and Lactams in Chloroform, Journal of the American Chemical Society, *113(1)*, 209–216.

207. Allinger, N., Yuh, Y., & Lii, J. (1989). Molecular Mechanics, The MM3 Force Field for Hydrocarbons 1 Journal of the American Chemical Society, *111(23)*, 8551–8566.

208. Walther, J., et al. (2001). Carbon Nanotubes in Water Structural Characteristics and Energetics, The Journal of Physical Chemistry B, *105(41)*, 9980–9987.

209. Robertson, D., Brenner, D., & Mintmire, J. (1992). Energetics of Nanoscale Graphitic Tubules, Physical Review B, *45(21)*, 12592.

210. Odegard, G. et al. (2002). Equivalent-Continuum Modeling of Nano-Structured Materials, Composites Science and Technology, *62(14)*, 1869–1880.

211. Odegard, G. et al. (2001). Equivalent-Continuum Modeling of Nano-Structured Materials.

212. Machida, K. (1999). Principles of Molecular Mechanics, Wiley 544.

213. Rappé, A. et al. (1992). UFF, a Full Periodic Table Force Field for Molecular Mechanics and Molecular Dynamics Simulations, Journal of the American Chemical Society, *114(25)*, 10024–10035.

214. Mayo, S., Olafson, B., & Goddard, W. (1990). DREIDING a Generic Force Field for Molecular Simulations, Journal of Physical Chemistry, *94(26)*, 8897–8909.

215. Kelly, B. (1981). Physics of Graphite, *3*, Applied Science London, 237.

216. Kudin, K., Scuseria, G., & Yakobson, B. (2001). C 2 F, BN, and C Nanoshell Elasticity from Ab Initio Computations, Physical Review B, *64(23)*, 235406.

217. Jorgensen, W., & Severance, D. (1990). Aromatic-Aromatic Interactions Free Energy Profiles for the Benzene Dimer in Water, Chloroform, and Liquid Benzene, Journal of the American Chemical Society, *112(12)*, 4768–4774.

218. Heermann, D. (1990). Computer-Simulation Methods, Springer.

219. Haile, J. (1992). Molecular Dynamics Simulation, Elementary Methods, John Wiley & Sons, Inc.

220. Sohlberg, K. et al. (1998). Continuum Methods of Mechanics as a Simplified Approach to Structural Engineering of Nanostructures, Nanotechnology, *9(1)*, 30.

221. Kresin, V., & Aharony, A. (1995). Fully Collapsed Carbon Nanotubes, Nature, *377*, 135.

222. Zhong-Can, O., Su, Z., & Wang, C. (1997). Coil Formation in Multishell Carbon Nanotubes, Competition between Curvature Elasticity and Interlayer Adhesion, Physical Review Letters, *78(21)*, 4055.

223. Yang, Y., Tobias, I., & Olson, W. (1993). Finite Element Analysis of DNA Supercoiling, The Journal of Chemical Physics, *98(2)*, 1673–1686.

224. Yu, M., Kowalewski, T., & Ruoff, R. (2001). Structural Analysis of Collapsed, and Twisted and Collapsed, Multi-walled Carbon Nanotubes by Atomic Force Microscopy, Physical Review Letters, *86(1)*, 87–93.

225. Avriel, M. (2012). Nonlinear Programming, Analysis and Methods, Courier Dover Publications.

226. Snyman, J. (2005). Practical Mathematical Optimization an Introduction to Basic Optimization Theory and Classical and New Gradient-Based Algorithms, *97*, Springer.

227. Atkinson, K. (2008). An Introduction to Numerical Analysis, John Wiley & Sons.

228. Golub, G., & Van Loan, C. (2012). Matrix Computations, *3*, JHU Press 422.

229. Morse, P. (1929). Diatomic Molecules According to the Wave Mechanics II, Vibrational Levels, Physical Review, *34(1)*, 57–64.

230. Barron, T., & Domb, C. (1955). On the Cubic and Hexagonal Close-Packed Lattices, Proceedings of the Royal Society of London Series A Mathematical and Physical Sciences, *227(1171)*, 447–465.

231. Wilson, S., Bernath, P., & McWeeny, R. (2003). Handbook of Molecular Physics and Quantum Chemistry, *2*, Wiley Chichester.

232. Tersoff, J. (1988). New Empirical Approach for the Structure and Energy of Covalent Systems, Physical Review B, *37(12)*, 69–91.

233. Brenner, D. (1990). Empirical Potential for Hydrocarbons for Use in Simulating the Chemical Vapor Deposition of Diamond Films, Physical Review B, *42(15)*, 9458–9466.

234. Allinger, N., & Sprague, J. (1973). Conformational Analysis XC Calculation of the Structures of Hydrocarbons Containing Delocalized Electronic Systems by the Molecular Mechanics Method, Journal of the American Chemical Society, *95(12)*, 3893–3907.

235. Weiner, P., & Kollman, P. (1981). AMBER Assisted Model Building with Energy Refinement, A General Program for Modeling Molecules and their Interactions, Journal of Computational Chemistry, *2(3)*, 287–303.

236. Brooks, B. et al. (1983). CHARMM A Program for Macromolecular Energy, Minimization, and Dynamics Calculations, Journal of Computational Chemistry, *4(2)*, 187–217.

237. Van Gunsteren, W. et al. (1983). Computer Simulation of the Dynamics of Hydrated Protein Crystals and its Comparison with X-ray Data, Proceedings of the National Academy of Sciences, *80(14)*, 4315–4319.

238. Caillerie, D., Mourad, A., & Raoult, A. (2006). Discrete Homogenization in Graphene Sheet Modeling, Journal of Elasticity, *84(1)*, 33–68.

239. Arroyo, M., & Belytschko, T. (2004). Finite Crystal Elasticity of Carbon Nanotubes Based on the Exponential Cauchy-Born Rule, Physical Review B, *69(11)*, 115–122.

240. Arroyo, M., & Belytschko, T. (2004). Finite Element Methods for the Non☐Linear Mechanics of Crystalline Sheets and Nanotubes, International Journal for Numerical Methods in Engineering, *59(3)*, 419–456.

241. Friesecke, G., & Theil, F. (2002). Validity and Failure of the Cauchy-Born Hypothesis in a Two-Dimensional Mass-Spring Lattice, Journal of Nonlinear Science, *12(5)*, 445–478.

242. Malvern, L. (1969). Introduction to the Mechanics of a Continuous Medium, 300.

243. Cousins, C. (1978). Inner Elasticity, Journal of Physics C Solid State Physics, *11(24)*, 486–499.

244. Martin, J. (1975). Many-Body Forces in Metals and the Brugger Elastic Constants, Journal of Physics C Solid State Physics, *8(18)*, 28–37.

245. Ericksen, J. (2005). Nonlinear Elasticity of Diatomic Crystals, Mechanics and Mathematics of Crystals, Selected Papers of JL Ericksen, 7–16.

246. Tadmor, E. et al. (1999). Mixed Finite Element and Atomistic Formulation for Complex Crystals, Physical Review B, *59(1)*, 235–248.

247. Marsden, J., & Hughes, T. (1983). Mathematical Foundations of Elasticity (1994) Prentice-Hall, Englewood Cliffs, NJ.

248. Do Carmo, M. (1976). Differential Geometry of Curves and Surfaces, *2*, Prentice-Hall Englewood Cliffs.

249. Morgan, F., & Bredt, J. (1998). Riemannian Geometry a Beginner's Guide, AK Peters Wellesley, 560.
250. Arroyo, M., & Belytschko, T. (2004). Finite Crystal Elasticity of Carbon Nanotubes Based on the Exponential Cauchy-Born Rule, Physical Review B, *69(11)*, 115–129.
251. Ericksen, J. (2008). On the Cauchy-Born Rule, Mathematics and Mechanics of Solids, *13(3–4)*, 199–220.
252. Cousins, C. (1978). Inner Elasticity, Journal of Physics C, Solid State Physics, *11(24)*, 48–67.
253. Brenner, D. et al. (2002). A Second-Generation Reactive Empirical Bond Order (REBO) Potential Energy Expression for Hydrocarbons, Journal of Physics Condensed Matter, *14(4)*, 783.
254. White, C. T., Robertson, D. H., & Mintmire, J. W. (1993). Helical and Rotational Symmetries of Nanoscale Graphitic Tubules Physical Review B, *47*, 5485–5488.
255. Milosevic, I., & Damnjanovic, M. (1993). Normal Vibrations and Jahn-Teller Effect for Polymers and Quasi-One-Dimensional Systems, Physical Review, *47*, 7805–7818.
256. Milosevic, I., Zivanovic, R., & Damnjanovic, M. (1997). Symmetry Classification of Stereoregular Polymers, Polymer, *38*, 4445–4453.
257. Caillerie, D., Mourad, A., & Raoult, A. (2006). Discrete Homogenization in Graphene Sheet Modeling Journal of Elasticity, *84*, 33–68.
258. Damnjanovic, M. et al. (1999). Symmetry and Lattices of Single-Wall Nanotubes, Journal of Physics A, *32*, 4097–4104.
259. Allinger, N. L. (1977). Conformational Analysis 130.MM2, A Hydrocarbon Force Field Utilizing V1 and V2 Torsional Terms, Journal of the American Chemical Society, *99*, 8127–8134
260. Odegard, G. et al. (2001). Equivalent-Continuum Modeling of Nano-Structured Materials, 17–28.
261. Liu, B. et al. (2004). The Atomic-Scale Finite Element Method, Computational Methods in Applied Mechanics and Engineering, *193*, 1849–1864.
262. Carroll, C. P. et al. (2008). Nanofibers from Electrically Driven Viscoelastic Jets, Modeling and Experiments Korea-Australia Rheology Journal, *20(3)*, 153–164.
263. Press, W. et al. (1990). Numerical Recipes Cambridge University Press Cambridge, 56–66.
264. Liu, B. et al. (2005). Atomic-Scale Finite Element Method in Multiscale Computation with Applications to Carbon Nanotubes, Physical Review B, *72(3)*, 422–435.
265. Rahman, A. (1964). Correlations in the Motion of Atoms in Liquid Argon, Physical Review, *136(2A)*, 405–412.
266. Pantano, A. D., Parks, M., & Boyce, M. (2004). Mechanics of Deformation of Single- and Multi-wall Carbon Nanotubes, Journal of the Mechanics and Physics of Solids, *52(4)*, 789–821.
267. Zhou, X. et al. (2001). The Structure Relaxation of Carbon Nanotube, Physica B Condensed Matter, *304(1)*, 86–90.
268. Peralta-Inga, Z. et al. (2003). Density Functional Tight-Binding Studies of Carbon Nanotube Structures, Structural Chemistry, *14(5)*, 431–443.
269. Salvetat, J. et al. (1999). Mechanical Properties of Carbon Nanotubes, Applied Physics A, *69(3)*, 255–260.

270. Poncharal, P. et al. (1999). Electrostatic Deflections and Electromechanical Resonances of Carbon Nanotubes, Science, *283(5407)*, 1513–1516.
271. Liu, J., Zheng, Q., & Jiang, Q. (2001). Effect of a Rippling Mode on Resonances of Carbon Nanotubes, Physical Review Letters, *86(21)*, 4843.
272. Tu, Z., & Ou-Yang, Z. (2002). Single-Walled and Multi-walled Carbon Nanotubes Viewed as Elastic Tubes with the Effective Young's Moduli Dependent on Layer Number, Physical Review B, *65(23)*, 233–245.
273. Chang, T., Geng, J., & Guo, X. (2005). Chirality-and Size-Dependent Elastic Properties of Single-Walled Carbon Nanotubes, Applied Physics Letters, *87(25)*, 251–277.
274. Goze, C. et al. (1999). Elastic and Mechanical Properties of Carbon Nanotubes, Synthetic Metals, *103(1)*, 2500–2501.
275. Yakobson, B. (1998). Mechanical Relaxation and "Intramolecular Plasticity" in Carbon Nanotubes, Applied Physics Letters, *72(8)*, 918–920.
276. Wei, C., Cho, K., & Srivastava, D. (2003). Tensile Yielding of Multi-wall Carbon Nanotubes, Applied Physics Letters, *82(15)*, 2512–2514.
277. Frackowiak, E. et al. (1999). Electrochemical Storage of Lithium in Multi-walled Carbon Nanotubes, Carbon, *37(1)*, 61–69.
278. Wu, G. et al. (1999). Structure and Lithium Insertion Properties of Carbon Nanotubes, Journal of the Electrochemical Society, *146(5)*, 1696–1701.
279. Claye, A. et al. (2000). Solid-State Electrochemistry of the Li Single Wall Carbon Nanotube System, Journal of the Electrochemical Society, *147(8)*, 2845–2852.
280. Shimoda, H. et al. (2002). Lithium Intercalation into Etched Single-Wall Carbon Nanotubes, Physica B Condensed Matter, *323(1)*, 133–134.
281. Gao, B. et al. (1999). Electrochemical Intercalation of Single-Walled Carbon Nanotubes with Lithium, Chemical Physics Letters, *307(3)*, 153–157.
282. Kong, J. et al. (2000). Nanotube Molecular Wires as Chemical Sensors, Science, *287(5453)*, 622–625.
283. Duesberg, G. et al. (1999). Towards Processing of Carbon Nanotubes for Technical Applications, Applied Physics A, *69(3)*, 269–274.
284. Pederson, M., & Broughton, J. (1992). Nanocapillarity in Fullerene Tubules, Physical Review Letters, *69(18)*, 2689.
285. Dujardin, E. et al. (1994). Capillarity and Wetting of Carbon Nanotubes, Science, *265(5180)*, 1850–1852.
286. Ren, X. et al. (2011). Carbon Nanotubes as Adsorbents in Environmental Pollution Management, a Review Chemical Engineering Journal, *170(2)*, 395–410.
287. Chen, W., Duan, L., & Zhu, D. (2007). Adsorption of Polar and Nonpolar Organic Chemicals to Carbon Nanotubes, Environmental Science & Technology, *41(24)*, 8295–8300.
288. Gotovac, S. et al. (2007). Assembly Structure Control of Single Wall Carbon Nanotubes with Liquid Phase Naphthalene Adsorption, Colloids and Surfaces A Physicochemical and Engineering Aspects, *300(1)*, 117–121.
289. Gotovac, S. et al. (2007). Effect of Nanoscale Curvature of Single-Walled Carbon Nanotubes on Adsorption of Polycyclic Aromatic Hydrocarbons, Nano Letters, *7(3)*, 583–587.

290. Yang, K., Zhu, L., & Xing, B. (2006). Adsorption of Polycyclic Aromatic Hydrocarbons by Carbon Nanomaterials, Environmental Science & Technology, *40(6)*, 1855–1861.
291. Jiao, L. et al. (2009). Narrow Graphene Nanoribbons from Carbon Nanotubes, Nature, *458(7240)*, 877–880.
292. Liu, T. et al. (2012). Adsorption of Methylene Blue from Aqueous Solution by Graphene, Colloids and Surfaces B Biointerfaces, *90*, 197–203.
293. Lin, D., & Xing, B. (2008). Adsorption of Phenolic Compounds by Carbon Nanotubes, Role of Aromaticity and Substitution of Hydroxyl Groups, Environmental Science & Technology, *42(19)*, 7254–7259.
294. Peng, X. et al. (2003). Adsorption of 1, 2-Dichlorobenzene from Water to Carbon Nanotubes, Chemical Physics Letters, *376(1)*, 154–158.
295. Rosenzweig, S. et al. (2013). Effect of Acid and Alcohol Network Forces within Functionalized Multi-wall Carbon Nanotubes Bundles on Adsorption of Copper (II) Species, Chemosphere, *90(2)*, 395–402.
296. Cho, H. et al. (2008). Influence of Surface Oxides on the Adsorption of Naphthalene onto Multi-walled Carbon Nanotubes, Environmental Science & Technology, *42(8)*, 2899–2905.
297. Wu, W. et al. (2012). Influence of pH and Surface Oxygen-Containing Groups on Multi-walled Carbon Nanotubes on the Transformation and Adsorption of 1-Naphthol, Journal of Colloid and Interface Science, *374(1)*, 226–231.
298. Lu, C., Chung, Y., & Chang, K. (2006). Adsorption Thermodynamic and Kinetic Studies of Trihalomethanes on Multi-walled Carbon Nanotubes, Journal of Hazardous Materials, *138(2)*, 304–310.
299. Sheng, G. et al. (2010). Adsorption of Copper (II) on Multi-walled Carbon Nanotubes in the Absence and Presence of Humic or Fulvic Acids, Journal of Hazardous Materials, *178(1)*, 333–340.
300. Yao, Y. et al. (2011). Equilibrium and Kinetic Studies of Methyl Orange Adsorption on Multi-walled Carbon Nanotubes, Chemical Engineering Journal, *170(1)*, 82–89.
301. Lu, C., Chung, Y., & Chang, K. (2005). Adsorption of Trihalomethanes from Water with Carbon Nanotubes, Water Research, *39(6)*, 1183–1189.
302. Ji, L. et al. (2009). Mechanisms for Strong Adsorption of Tetracycline to Carbon Nanotubes, A Comparative Study Using Activated Carbon and Graphite as Adsorbents, Environmental Science & Technology, *43(7)*, 2322–2327.
303. Zhang, S. et al. (2012). Adsorption Kinetics of Aromatic Compounds on Carbon Nanotubes and Activated Carbons, Environmental Toxicology and Chemistry, *31(1)*, 79–85.
304. Chen, J., Chen, W., & Zhu, D. (2008). Adsorption of Nonionic Aromatic Compounds to Single-Walled Carbon Nanotubes, Effects of Aqueous Solution Chemistry, Environmental Science & Technology, *42(19)*, 7225–7230.
305. Zhang, S. et al. (2010). Adsorption of Synthetic Organic Chemicals by Carbon Nanotubes, Effects of Background Solution Chemistry, Water Research, *44(6)*, 2067–2074.
306. Wang, X., Lu, J., & Xing, B. (2008). Sorption of Organic Contaminants by Carbon Nanotubes, Influence of Adsorbed Organic Matter, Environmental Science & Technology, *42(9)*, 3207–3212.

307. Pan, B. et al. (2008). Adsorption and Hysteresis of Bisphenol A and 17α-Ethinyl Estradiol on Carbon Nanomaterials, Environmental Science & Technology, *42(15)*, 5480–5485.

308. Woods, L., Bădescu, Ş., & Reinecke, T. (2007). Adsorption of Simple Benzene Derivatives on Carbon Nanotubes, Physical Review B, *75(15)*, 155415.

309. Tournus, F., & Charlier, J. (2005). Ab Initio Study of Benzene Adsorption on Carbon Nanotubes, Physical Review B, *71(16)*, 165421.

310. Zou, M. et al. (2012). Simulating Adsorption of Organic Pollutants on Finite (8, 0) Single-Walled Carbon Nanotubes in Water, Environmental Science & Technology, *46(16)*, 8887–8894.

311. Gui, X. et al. (2010). Carbon Nanotube Sponges, Advanced Materials, *22(5)*, 617–621.

312. Chen, C., Wang, X., & Nagatsu, M. (2009). Europium Adsorption on Multi-wall carbon Nanotube/Iron Oxide Magnetic Composite in the Presence of Polyacrylic Acid, Environmental Science & Technology, *43(7)*, 2362–2367.

313. Zhu, H. et al. (2010). Preparation, Characterization, Adsorption Kinetics and Thermodynamics of Novel Magnetic Chitosan Enwrapping Nanosized γ-Fe$_2$O$_3$ and Multi-walled Carbon Nanotubes with Enhanced Adsorption Properties for Methyl Orange, Bioresource Technology, *101(14)*, 5063–5069.

314. Zeng, Y. et al. (2013). Enhanced Adsorption of Malachite Green onto Carbon Nanotube/Polyaniline Composites, Journal of Applied Polymer Science, *127(4)*, 2475–2482.

315. Di, Z. et al. (2006). Chromium Adsorption by Aligned Carbon Nanotubes Supported Ceria Nanoparticles, Chemosphere, *62(5)*, 861–865.

316. Maggini, L. et al. (2013). Magnetic Poly (Vinylpyridine)-Coated Carbon Nanotubes, An Efficient Supramolecular Tool for Wastewater Purification, ChemSusChem, *6(2)*, 367–373.

317. Li, J. et al. (2011). Effect of Surfactants on Pb (II) Adsorption from Aqueous Solutions Using Oxidized Multi-wall Carbon Nanotubes, Chemical Engineering Journal, *166(2)*, 551–558.

318. Dresselhaus, M., Dresselhaus, G., & Eklund, P.C. (1996). Science of Fullerenes and Carbon Nanotubes, their Properties and Applications, Academic Press.

319. Dresselhaus, M. et al. (1988). Synthesis of Graphite Fibers and Filaments, in Graphite Fibers and Filaments, Springer, 12–34.

320. Lieber, C. (1998). One-Dimensional Nanostructures, Chemistry, Physics & Applications, Solid State Communications, *107(11)*, 607–616.

321. Dresselhaus, M., Williams, K., & Eklund, P. (1999). Hydrogen Adsorption in Carbon Materials, Mrs Bulletin, *24(11)*, 45–50.

322. Ajayan, P. (1999). Nanotubes from Carbon, Chemical Reviews, *99(7)*, 1787–1800.

323. Nardelli, M., Yakobson, B., & Bernholc, J. (1998). Brittle and Ductile Behavior in Carbon Nanotubes, Physical Review Letters, *81(21)*, 4656.

324. Stone, A., & Wales, D. (1986). Theoretical Studies of Icosahedral C$_{60}$ and Some Related Species, Chemical Physics Letters, *128(5)*, 501–503.

325. Calvert, P. (1999). Nanotube Composites, a Recipe for Strength, Nature, *399(6733)*, 210–211.

326. Ajayan, P., & Zhou, O. (2001). Applications of Carbon Nanotubes, in Carbon Nanotubes, Springer, 391–425.
327. Curran, S. et al. (1998). A Composite from Poly (m-phenylenevinylene-co-2, 5-dioctoxy-p-phenylenevinylene) and Carbon Nanotubes, A Novel Material for Molecular Optoelectronics, Advanced Materials, *10(14)*, 1091–1093.
328. Coleman, J. et al. (2000). Phase Separation of Carbon Nanotubes and Turbostratic Graphite Using a Functional Organic Polymer, Advanced Materials, *12(3)*, 213–216.
329. Ago, H. et al. (1999). Composites of Carbon Nanotubes and Conjugated Polymers for Photovoltaic Devices, Advanced Materials, *11(15)*, 1281–1285.
330. Carroll, D. et al. (1997). Electronic Structure and Localized States at Carbon Nanotube Tips, Physical Review Letters, *78(14)*, 2811.
331. Jones, A., Bekkedahl, T., & Kiang, C. (1997). Storage of Hydrogen in Single-Walled Carbon Nanotubes, Nature, 386, 377.
332. Chen, P. et al. (1999). High H2 Uptake by Alkali-Doped Carbon Nanotubes Under Ambient Pressure and Moderate Temperatures, Science, *285(5424)*, 91–93.
333. Liu, C. et al. (1999). Hydrogen Storage in Single-Walled Carbon Nanotubes at Room Temperature, Science, *286(5442)*, 1127–1129.
334. Vaccari, L. et al. (2007). Carbon Nanotubes, in Nanostructures-Fabrication and Analysis, Springer, 151–215.
335. Chambers, A. et al. (1998). Hydrogen Storage in Graphite Nanofibers, The Journal of Physical Chemistry B, *102(22)*, 4253–4256.
336. Cheng, H., Yang, Q., & Liu, C. (2001). Hydrogen Storage in Carbon Nanotubes, Carbon, *39(10)*, 1447–1454.
337. Ye, Y. et al. (1999). Hydrogen Adsorption and Cohesive Energy of Single-Walled Carbon Nanotubes, Applied Physics Letters, *74(16)*, 2307–2309.
338. Collins, P. et al. (2000). Extreme Oxygen Sensitivity of Electronic Properties of Carbon Nanotubes, Science, *287(5459)*, 1801–1804.
339. Tang, X. et al. (2000). Electronic Structures of Single-Walled Carbon Nanotubes Determined by NMR Science, *288(5465)*, 492–494.
340. He, Z. et al. (2011). Etchant-Induced Shaping of Nanoparticle Catalysts During Chemical Vapour Growth of Carbon Nanofibers, Carbon, *49(2)*, 435–444.
341. Bárdos, L., & Baránková, H. (2010). Cold Atmospheric Plasma Sources, Processes, and Applications, Thin Solid Films, *518(23)*, 6705–6713.
342. Unrau, C., Axelbaum, R., & Lo, C. (2010). High-Yield Growth of Carbon Nanotubes on Composite Fe/Si/O Nanoparticle Catalysts, A Car-Parrinello Molecular Dynamics and Experimental Study, The Journal of Physical Chemistry C, *114(23)*, 10430–10435.
343. Kruusenberg, I. et al. (2011). Effect of Purification of Carbon Nanotubes on their Electrocatalytic Properties for Oxygen Reduction in Acid Solution, Carbon, *49(12)*, 4031–4039.
344. Ebbesen, T., & Ajayan, P. (1992). Large-Scale Synthesis of Carbon Nanotubes, Nature, *358(6383)*, 220–222.
345. Sakurai, T. et al. (1996). Scanning Tunneling Microscopy Study of Fullerenes, Progress in Surface Science, *51(4)*, 263–408.
346. Zhao, X. et al. (1996). Morphology of Carbon Nanotubes Prepared by Carbon Arc, Japanese Journal of Applied Physics, *35(*part 1), 4451–4456.

347. Zhao, X. et al. (1997). Preparation of High-Grade Carbon Nanotubes by Hydrogen Arc Discharge, Carbon, *35(6)*, 775–781.
348. Zhao, X. et al. (1999). Morphology of Carbon Allotropes Prepared by Hydrogen Arc Discharge, Journal of Crystal Growth, *198*, 934–938.
349. Anazawa, K. et al. (2002). High-Purity Carbon Nanotubes Synthesis Method by an Arc Discharging in Magnetic Field, Applied Physics Letters, *81(4)*, 739–741.
350. Jiang, Y. et al. (2009). Influence of NH3 Atmosphere on the Growth and Structures of Carbon Nanotubes Synthesized by the Arc-Discharge Method, Inorganic Materials, *45(11)*, 1237–1239.
351. Parkansky, N. et al. (2004). Single-Pulse Arc Production of Carbon Nanotubes in Ambient Air, Journal of Physics D Applied Physics, *37(19)*, 2715.
352. Jung, S. et al. (2003). High-Yield Synthesis of Multi-walled Carbon Nanotubes by Arc Discharge in Liquid Nitrogen, Applied Physics A, *76(2)*, 285–286.
353. Prasek, J. et al. (2011). Methods for Carbon Nanotubes Synthesis-Review Journal of Materials Chemistry, *21(40)*, 15872–15884.
354. Montoro, L., Lofrano, R., & Rosolen, J. (2005). Synthesis of Single-Walled and Multi-walled Carbon Nanotubes by Arc-Water Method, Carbon, *43(1)*, 200–203.
355. Guo, J. et al. (2007). Structure of Nanocarbons Prepared by Arc Discharge in Water, Materials Chemistry and Physics, *105(2)*, 175–178.
356. Xing, G., Jia, S., & Shi, Z. (2007). The Production of Carbon Nano-Materials by Arc Discharge Under Water or Liquid Nitrogen, New Carbon Materials, *22(4)*, 337–341.
357. Iijima, S., & Ichihashi, T. (1993). Single-Shell Carbon Nanotubes of 1 nm Diameter, 8–15.
358. Seraphin, S. (1995). Single-Walled Tubes and Encapsulation of Nanocrystals into Carbon Clusters, Journal of the Electrochemical Society, *142(1)*, 290–297.
359. Tomita, M., & Hayashi, T. (1994). Single-Wall Carbon Nanotubes Growing Radially From Ni Fine Particles Formed by Arc Evaporation, Japan J Appl Phys, *(2)*, 4–11.
360. Saito, Y. et al. (1996). Carbon Nanocapsules and Single-Layered Nanotubes Produced with Platinum-Group Metals (Ru, Rh, Pd, Os, Ir, Pt) by Arc Discharge, Journal of Applied Physics, *80(5)*, 3062–3067.
361. Chen, B. et al. (2010). Fabrication and Dispersion Evaluation of Single-Wall Carbon Nanotubes Produced by FH-Arc Discharge Method, Journal of Nanoscience and Nanotechnology, *10(6)*, 3973–3977.
362. Chen, B., Inoue, S., & Ando, Y. (2009). Raman Spectroscopic and Thermogravimetric Studies of High-Crystallinity SWNTs Synthesized by FH-Arc Discharge Method, Diamond and Related Materials, *18(5)*, 975–978.
363. Zhao, J., Bao, W., & Liu, X. (2010). Synthesis of SWNTs from Charcoal by Arc-Discharging, Journal of Wuhan University of Technology-Mater, Science Ed., *25(2)*, 194–196.
364. Wang, H. et al. (2010). Influence of Mo on the Growth of Single-Walled Carbon Nanotubes in Arc Discharge, Journal of Nanoscience and Nanotechnology, *10(6)*, 3988–3993.
365. Liang, C., Li, Z., & Dai, S. (2008). Mesoporous Carbon Materials, Synthesis and Modification, Angewandte Chemie International Edition, *47(20)*, 3696–3717.
366. Hutchison, J. et al. (2001). Double-Walled Carbon Nanotubes Fabricated by a Hydrogen Arc Discharge Method, Carbon, *39(5)*, 761–770.

367. Sugai, T. et al. (2003). New Synthesis of High-Quality Double-Walled Carbon Nanotubes by High-Temperature Pulsed Arc Discharge, Nano Letters, *3(6)*, 769–773.

368. Huang, H. et al. (2003). High-Quality Double-Walled Carbon Nanotube Super Bundles Grown in a Hydrogen-Free Atmosphere, The Journal of Physical Chemistry B, *107(34)*, 8794–8798.

369. Qiu, J. et al. (2007). Synthesis of Double-Walled Carbon Nanotubes from Coal in Hydrogen-Free Atmosphere, Fuel, *86(1)*, 282–286.

370. Qiu, H. et al. (2006). Synthesis and Raman Scattering Study of Double-Walled Carbon Nanotube Peapods, Solid State Communications, *137(12)*, 654–657.

371. Liu, Q. et al. (2009). Semiconducting Properties of Cup-Stacked Carbon Nanotubes, Carbon, *47(3)*, 731–736.

372. Liu, C., & Cheng, H. (2005). Carbon Nanotubes for Clean Energy Applications, Journal of Physics D Applied Physics, *38(14)*, R231.

373. Ando, Y. et al. (2000). Multi-walled Carbon Nanotubes Prepared by Hydrogen Arc, Diamond and Related Materials, *9(3)*, 847–851.

374. Pillai, S., et al. (2008). The Effect of Calcination on Multi-walled Carbon Nanotubes Produced by Dc-Arc Discharge. Journal of nanoscience and nanotechnology, *8(7)*, 3539–3544.

375. Dunens, O., MacKenzie, K., & Harris, A. (2010). Large-Scale Synthesis of Double-Walled Carbon Nanotubes in Fluidized Beds, Industrial & Engineering Chemistry Research, *49(9)*, 4031–4035.

376. Ikegami, T. et al. (2004). Optical Measurement in Carbon Nanotubes Formation by Pulsed Laser Ablation, Thin Solid Films, *457(1)*, 7–11.

377. Bolshakov, A. et al. (2002). A Novel CW Laser-Powder Method of Carbon Single-Wall Nanotubes Production, Diamond and Related Materials, *11(3)*, 927–930.

378. Zhang, M. et al. (2005). Strong, Transparent, Multifunctional, Carbon Nanotube Sheets, Science, *309(5738)*, 1215–1219.

379. Lebel, L. et al. (2010). Preparation and Mechanical Characterization of Laser Ablated Single-Walled Carbon-Nanotubes/Polyurethane Nanocomposite Microbeams, Composites Science and Technology, *70(3)*, 518–524.

380. Kusaba, M., & Tsunawaki, Y. (2006). Production of Single-Wall Carbon Nanotubes by a XeCl Excimer Laser Ablation, Thin Solid Films, *506*, 255–258.

381. Stramel, A. et al. (2010). Pulsed Laser Deposition of Carbon Nanotube and Polystyrene-Carbon Nanotube Composite Thin Films, Optics and Lasers in Engineering, *48(12)*, 1291–1295.

382. Bonaccorso, F. et al. (2007). Pulsed Laser Deposition of Multi-walled Carbon Nanotubes Thin Films, Applied Surface Science, *254(4)*, 1260–1263.

383. Steiner, III S. et al. (2009). Nanoscale Zirconia as a Nonmetallic Catalyst for Graphitization of Carbon and Growth of Single-and Multi-wall Carbon Nanotubes, Journal of the American Chemical Society, *131(34)*, 12144–12154.

384. Tempel, H., Joshi, R., & Schneider, J. (2010). Ink Jet Printing of Ferritin as Method for Selective Catalyst Patterning and Growth of Multi-walled Carbon Nanotubes, Materials Chemistry and Physics, *121(1)*, 178–183.

385. Smajda, R. et al. (2009). Synthesis and Mechanical Properties of Carbon Nanotubes Produced by the Water Assisted CVD Process, Physica Status Solidi (b), *246(11, 12)*, 2457–2460.

386. Byon, H. et al. (2007). A Synthesis of High Purity Single-Walled Carbon Nanotubes from Small Diameters of Cobalt Nanoparticles by Using Oxygen-Assisted Chemical Vapor Deposition Process, BULLETIN-KOREAN CHEMICAL SOCIETY, *28(11)*, 2056.

387. Varshney, D., Weiner, B. R., & Morell, G. (2010). Growth and Field Emission Study of a Monolithic Carbon Nanotube/Diamond Composite, Carbon,. *48(12)*, 3353–3358.

388. Chatrchyan, S. et al. (2011). Search for Supersymmetry at the LHC in Events with Jets and Missing Transverse Energy, Physical Review Letters, *107(22)*, 221804.

389. Brown, B. et al. (2011). Growth of Vertically Aligned Bamboo-Like Carbon Nanotubes from Ammonia/Methane Precursors Using a Platinum Catalyst, Carbon, *49(1)*, 266–274.

390. Xu, Y. et al. (2011). Chirality-Enriched Semiconducting Carbon Nanotubes Synthesized on High Surface Area MgO-Supported Catalyst, Materials Letters, *65(12)*, 1878–1881.

391. Wang, W., Zhu, Y., & Yang, L. (2007). ZnO–SnO2 Hollow Spheres and Hierarchical Nanosheets, Hydrothermal Preparation, Formation Mechanism, and Photocatalytic Properties, Advanced Functional Materials, *17(1)*, 59–64.

392. Fotopoulos, N., & Xanthakis, J. (2010). A Molecular Level Model for the Nucleation of a Single-Wall Carbon Nanotube Cap Over a Transition Metal Catalytic Particle, Diamond and Related Materials, *19(5)*, 557–561.

393. Sharma, R. et al. (2011). Evaluation of the Role of Au in Improving Catalytic Activity of Ni Nanoparticles for the Formation of One-Dimensional Carbon Nanostructures, Nano Letters, *11(6)*, 2464–2471.

394. Palizdar, M. et al. (2011). Investigation of Fe/MgO Catalyst Support Precursors for the Chemical Vapour Deposition Growth of Carbon Nanotubes, Journal of Nanoscience and Nanotechnology, *11(6)*, 5345–5351.

395. Tomie, T. et al. (2010). Prospective Growth Region for Chemical Vapor Deposition Synthesis of Carbon Nanotube on C–H–O Ternary Diagram, Diamond and Related Materials, *19(11)*, 1401–1404.

396. Narkiewicz, U. et al. (2010). Catalytic Decomposition of Hydrocarbons on Cobalt, Nickel and Iron Catalysts to Obtain Carbon Nanomaterials, Applied Catalysis A General, *384(1)*, 27–35.

397. He, D. et al. (2011). Growth of Carbon Nanotubes in Six Orthogonal Directions on Spherical Alumina Microparticles, Carbon, *49(7)*, 2273–2286.

398. Shukla, B. et al. (2010). Interdependency of Gas Phase Intermediates and Chemical Vapor Deposition Growth of Single Wall Carbon Nanotubes, Chemistry of Materials, *22(22)*, 6035–6043.

399. Afolabi, A. et al. (2011). Synthesis and Purification of Bimetallic Catalyzed Carbon Nanotubes in a Horizontal CVD Reactor, Journal of Experimental Nanoscience, *6(3)*, 248–262.

400. Zhu, J., Yudasaka, M., & Iijima S. (2003). A Catalytic Chemical Vapor Deposition Synthesis of Double-Walled Carbon Nanotubes Over Metal Catalysts Supported on a Mesoporous Material, Chemical Physics Letters, *380(5)*, 496–502.

401. Ramesh, P. et al. (2005). Selective Chemical Vapor Deposition Synthesis of Double-Wall Carbon Nanotubes on Mesoporous Silica, The Journal of Physical Chemistry B, *109(3)*, 1141–1147.

402. Hiraoka, T. et al. (2003). Selective Synthesis of Double-Wall Carbon Nanotubes by CCVD of Acetylene Using Zeolite Supports, Chemical Physics Letters, *382(5)*, 679–685.

403. Flahaut, E., Laurent, C., & Peigney, A. (2005). Catalytic CVD Synthesis of Double and Triple-Walled Carbon Nanotubes by the Control of the Catalyst Preparation, Carbon, *43(2)*, 375–383.

404. Kim, S., & Gangloff, L. (2009). Growth of Carbon Nanotubes (CNTs) on Metallic Underlayers by Diffusion Plasma-Enhanced Chemical Vapour Deposition (DPECVD), Physica E Low-Dimensional Systems and Nanostructures, *41(10)*, 1763–1766.

405. Wang, H., & Moore, J. (2010). Different Growth Mechanisms of Vertical Carbon Nanotubes by rf-or dc-Plasma Enhanced Chemical Vapor Deposition at Low Temperature, Journal of Vacuum Science & Technology B, *28(6)*, 1081–1085.

406. Luais, E. et al. (2010). Preparation and Modification of Carbon Nanotubes Electrodes by Cold Plasmas Processes Toward the Preparation of Amperometric Biosensors, Electrochimica Acta, *55(27)*, 7916–7922.

407. Häffner, M. et al. (2010). Plasma Enhanced Chemical Vapor Deposition Grown Carbon Nanotubes from Ferritin Catalyst for Neural Stimulation Microelectrodes, Microelectronic Engineering, *87(5)*, 734–737.

408. Vollebregt, S. et al. (2010). Growth of High-Density Self-Aligned Carbon Nanotubes and Nanofibers Using Palladium Catalyst, Journal of Electronic Materials, *39(4)*, 371–375.

409. Duy, D. et al. (2009). Growth of Carbon Nanotubes on Stainless Steel Substrates by DC-PECVD, Applied Surface Science, *256(4)*, 1065–1068.

410. Vinten, P. et al. (2011). Origin of Periodic Rippling During Chemical Vapor Deposition Growth of Carbon Nanotube Forests, Carbon, *49(15)*, 4972–4981.

411. Kim, D. et al. (2008). Growth of Vertically Aligned Arrays of Carbon Nanotubes for High Field Emission, Thin Solid Films, *516(5)*, 706–709.

412. Kim, M. et al. (2003). Growth Characteristics of Carbon Nanotubes Via Aluminum Nanopore Template on Si Substrate Using PECVD, Thin Solid Films, *435(1)*, 312–317.

413. Byeon, H. et al. (2010). Growth of Ultra Long Multi-wall Carbon Nanotube Arrays by Aerosol-Assisted Chemical Vapor Deposition, Journal of Nanoscience and Nanotechnology, *10(9)*, 6116–6119.

414. Jeong, N. et al. (2013). Microscopic and Spectroscopic Analyzes of Pt-Decorated Carbon Nanowires Formed on Carbon Fiber Paper, Microscopy and Microanalysis, *19(S5)*, 198–201.

415. Liu, J. et al. (2011). Nitrogen-Doped Carbon Nanotubes with Tunable Structure and High Yield Produced by Ultrasonic Spray Pyrolysis, Applied Surface Science, *257(17)*, 7837–7844.

416. Khatri, I. et al. (2009). Synthesis of Single Walled Carbon Nanotubes by Ultrasonic Spray Pyrolysis Method, Diamond and Related Materials, *18(2)*, 319–323.

417. Camarena, J. et al. (2011). Molecular Assembly of Multi-wall Carbon Nanotubes with Amino Crown Ether, Synthesis and Characterization, Journal of Nanoscience and Nanotechnology, *11(6)*, 5539–5545.

418. Sadeghian, Z. (2009). Large-Scale Production of Multi-walled Carbon Nanotubes by Low-Cost Spray Pyrolysis of Hexane, New Carbon Materials, *24(1)*, 33–38.

419. Pinault, M. et al. (2004). Carbon Nanotubes Produced by Aerosol Pyrolysis, Growth Mechanisms and Post-Annealing Effects, Diamond and Related Materials, *13(4)*, 1266–1269.

420. Nebol'sin, V., & Vorob'ev, A. (2011). Role of Surface Energy in the Growth of Carbon Nanotubes via Catalytic Pyrolysis of Hydrocarbons, Inorganic Materials, *47(2)*, 128–132.

421. Lara-Romero, J. et al. (2011). Temperature Effect on the Synthesis of Multi-walled Carbon Nanotubes by Spray Pyrolysis of Botanical Carbon Feedstocks, Turpentine, α-pinene and β-pinene. Fullerenes, Nanotubes, and Carbon Nanostructures, *19(6)*, 483–496.

422. Kumar, R., Tiwari, R., & Srivastava, O. (2011). Scalable Synthesis of Aligned Carbon Nanotubes Bundles Using Green Natural Precursor, Neem Oil, Nanoscale Research Letters, *6(1)*, 1–6.

423. Paul, S., & Samdarshi, S. (2011). A Precursor for Carbon Nanotube Synthesis, New Carbon Materials, *26(2)*, 85–88.

424. Duan, W., & Wang, Q. (2010). Water Transport with a Carbon Nanotube Pump, ACS Nano, *4(4)*, 2338–2344.

425. Kucukayan, G. et al. (2011). An Experimental and Theoretical Examination of the Effect of Sulfur on the Pyrolytically Grown Carbon Nanotubes from Sucrose-Based Solid State Precursors, Carbon, *49(2)*, 508–517.

426. Clauss, C., Schwarz, M., & Kroke, E. (2010). Microwave-Induced Decomposition of Nitrogen-Rich Iron Salts and CNT Formation from Iron (III)-Melonate Fe Carbon, *48(4)*, 1137–1145.

427. Kuang, Z. et al. (2009). Biomimetic Chemosensor, Designing Peptide Recognition Elements for Surface Functionalization of Carbon Nanotube Field Effect Transistors, ACS Nano, *4(1)*, 452–458.

428. Du, G. et al. (2008). Solid-Phase Transformation of Glass-like Carbon Nanoparticles into Nanotubes and the Related Mechanism, Carbon, *46(1)*, 92–98.

429. El Hamaoui, B. et al. (2007). Solid-State Pyrolysis of Polyphenylene-Metal Complexes, A Facile Approach Toward Carbon Nanoparticles, Advanced Functional Materials, *17(7)*, 1179–1187.

430. Rudin, A., & Choi, P. (2012). The Elements of Polymer Science & Engineering, Academic Press.

431. Li, J. et al. (2003). Bottom-Up Approach for Carbon Nanotube Interconnects, Applied Physics Letters, *82(15)*, 2491–2493.

432. Jasti, R., & Bertozzi, C. (2010). Progress and Challenges for the Bottom-Up Synthesis of Carbon Nanotubes with Discrete Chirality, Chemical Physics Letters, *494(1)*, 1–7.

433. Omachi, H. et al. (2010). A Modular and Size-Selective Synthesis of [n] Cycloparaphenylenes A Step Toward the Selective Synthesis of [n, n] Single-Walled Carbon Nanotubes, Angewandte Chemie International Edition, *49(52)*, 10202–10205.

434. Segawa, Y. et al. (2011). Concise Synthesis and Crystal Structure of [12] Cycloparaphenylene, Angewandte Chemie International Edition, *50(14)*, 3244–3248.

435. Omachi, H., Segawa, Y., & Itami, K. (2011). Synthesis and Racemization Process of Chiral Carbon Nanorings, a Step Toward the Chemical Synthesis of Chiral Carbon Nanotubes, Organic Letters, *13(9)*, 2480–2483.

436. Iwamoto, T. et al. (2011). Selective and Random Syntheses of [n] Cycloparaphen-ylenes (n=8–13) and Size Dependence of their Electronic Properties, Journal of the American Chemical Society, *133(21)*, 8354–8361.
437. Fort, E., & Scott, L. (2011). Gas-Phase Diels-Alder Cycloaddition of Benzyne to an Aromatic Hydrocarbon Bay Region, Groundwork for the Selective Solvent-Free Growth of Armchair Carbon Nanotubes, Tetrahedron Letters, *52(17)*, 2051–2053.

INDEX

Milton Keynes UK
Ingram Content Group UK Ltd.
UKHW022104141024
449569UK00031B/1768

9 781774 631065